T0269072

LONDON MATHEMATICAL SOCIETY LECTURE NOTE SERIES

Managing Editor: Professor J.W.S. Cassels, Department of Pure Mathematics and Mathematical Statistics, University of Cambridge, 16 Mill Lane, Cambridge CB2 1SB, England

The titles below are available from booksellers, or, in case of difficulty, from Cambridge University Press.

34 Representation theory of Lie groups, M.F. ATIYAH *et al*
36 Homological group theory, C.T.C. WALL (ed)
39 Affine sets and affine groups, D.G. NORTHCOTT
46 p-adic analysis: a short course on recent work, N. KOBLITZ
50 Commutator calculus and groups of homotopy classes, H.J. BAUES
59 Applicable differential geometry, M. CRAMPIN & F.A.E. PIRANI
66 Several complex variables and complex manifolds II, M.J. FIELD
69 Representation theory, I.M. GELFAND *et al*
74 Symmetric designs: an algebraic approach, E.S. LANDER
76 Spectral theory of linear differential operators and comparison algebras, H.O. CORDES
77 Isolated singular points on complete intersections, E.J.N. LOOIJENGA
79 Probability, statistics and analysis, J.F.C. KINGMAN & G.E.H. REUTER (eds)
83 Homogeneous structures on Riemannian manifolds, F. TRICERRI & L. VANHECKE
86 Topological topics, I.M. JAMES (ed)
87 Surveys in set theory, A.R.D. MATHIAS (ed)
88 FPF ring theory, C. FAITH & S. PAGE
89 An F-space sampler, N.J. KALTON, N.T. PECK & J.W. ROBERTS
90 Polytopes and symmetry, S.A. ROBERTSON
91 Classgroups of group rings, M.J. TAYLOR
92 Representation of rings over skew fields, A.H. SCHOFIELD
93 Aspects of topology, I.M. JAMES & E.H. KRONHEIMER (eds)
94 Representations of general linear groups, G.D. JAMES
95 Low-dimensional topology 1982, R.A. FENN (ed)
96 Diophantine equations over function fields, R.C. MASON
97 Varieties of constructive mathematics, D.S. BRIDGES & F. RICHMAN
98 Localization in Noetherian rings, A.V. JATEGAONKAR
99 Methods of differential geometry in algebraic topology, M. KAROUBI & C. LERUSTE
100 Stopping time techniques for analysts and probabilists, L. EGGHE
101 Groups and geometry, ROGER C. LYNDON
103 Surveys in combinatorics 1985, I. ANDERSON (ed)
104 Elliptic structures on 3-manifolds, C.B. THOMAS
105 A local spectral theory for closed operators, I. ERDELYI & WANG SHENGWANG
106 Syzygies, E.G. EVANS & P. GRIFFITH
107 Compactification of Siegel moduli schemes, C-L. CHAI
108 Some topics in graph theory, H.P. YAP
109 Diophantine analysis, J. LOXTON & A. VAN DER POORTEN (eds)
110 An introduction to surreal numbers, H. GONSHOR
113 Lectures on the asymptotic theory of ideals, D. REES
114 Lectures on Bochner-Riesz means, K.M. DAVIS & Y-C. CHANG
115 An introduction to independence for analysts, H.G. DALES & W.H. WOODIN
116 Representations of algebras, P.J. WEBB (ed)
117 Homotopy theory, E. REES & J.D.S. JONES (eds)
118 Skew linear groups, M. SHIRVANI & B. WEHRFRITZ
119 Triangulated categories in the representation theory of finite-dimensional algebras, D. HAPPEL
121 Proceedings of *Groups - St Andrews 1985*, E. ROBERTSON & C. CAMPBELL (eds)
122 Non-classical continuum mechanics, R.J. KNOPS & A.A. LACEY (eds)
124 Lie groupoids and Lie algebroids in differential geometry, K. MACKENZIE
125 Commutator theory for congruence modular varieties, R. FREESE & R. MCKENZIE
126 Van der Corput's method of exponential sums, S.W. GRAHAM & G. KOLESNIK
127 New directions in dynamical systems, T.J. BEDFORD & J.W. SWIFT (eds)
128 Descriptive set theory and the structure of sets of uniqueness, A.S. KECHRIS & A. LOUVEAU
129 The subgroup structure of the finite classical groups, P.B. KLEIDMAN & M.W.LIEBECK
130 Model theory and modules, M. PREST
131 Algebraic, extremal & metric combinatorics, M-M. DEZA, P. FRANKL & I.G. ROSENBERG (eds)
132 Whitehead groups of finite groups, ROBERT OLIVER
133 Linear algebraic monoids, MOHAN S. PUTCHA
134 Number theory and dynamical systems, M. DODSON & J. VICKERS (eds)
135 Operator algebras and applications, 1, D. EVANS & M. TAKESAKI (eds)
136 Operator algebras and applications, 2, D. EVANS & M. TAKESAKI (eds)
137 Analysis at Urbana, I, E. BERKSON, T. PECK, & J. UHL (eds)

138   Analysis at Urbana, II,   E. BERKSON,   T. PECK, & J. UHL (eds)
139   Advances in homotopy theory,   S. SALAMON, B. STEER & W. SUTHERLAND (eds)
140   Geometric aspects of Banach spaces,   E.M. PEINADOR and A. RODES (eds)
141   Surveys in combinatorics 1989,   J. SIEMONS (ed)
142   The geometry of jet bundles,   D.J. SAUNDERS
143   The ergodic theory of discrete groups,   PETER J. NICHOLLS
144   Introduction to uniform spaces,   I.M. JAMES
145   Homological questions in local algebra,   JAN R. STROOKER
146   Cohen-Macaulay modules over Cohen-Macaulay rings,   Y. YOSHINO
147   Continuous and discrete modules,   S.H. MOHAMED & B.J. MÜLLER
148   Helices and vector bundles,   A.N. RUDAKOV *et al*
149   Solitons, nonlinear evolution equations and inverse scattering, M. ABLOWITZ & P. CLARKSON
150   Geometry of low-dimensional manifolds 1,   S. DONALDSON & C.B. THOMAS (eds)
151   Geometry of low-dimensional manifolds 2,   S. DONALDSON & C.B. THOMAS (eds)
152   Oligomorphic permutation groups,   P. CAMERON
153   L-functions and arithmetic,   J. COATES & M.J. TAYLOR (eds)
154   Number theory and cryptography,   J. LOXTON (ed)
155   Classification theories of polarized varieties,   TAKAO FUJITA
156   Twistors in mathematics and physics,   T.N. BAILEY & R.J. BASTON (eds)
157   Analytic pro-*p* groups,   J.D. DIXON, M.P.F. DU SAUTOY, A. MANN & D. SEGAL
158   Geometry of Banach spaces,   P.F.X. MÜLLER & W. SCHACHERMAYER (eds)
159   Groups St Andrews 1989 volume 1,   C.M. CAMPBELL & E.F. ROBERTSON (eds)
160   Groups St Andrews 1989 volume 2,   C.M. CAMPBELL & E.F. ROBERTSON (eds)
161   Lectures on block theory,   BURKHARD KÜLSHAMMER
162   Harmonic analysis and representation theory,   A. FIGA-TALAMANCA & C. NEBBIA
163   Topics in varieties of group representations,   S.M. VOVSI
164   Quasi-symmetric designs,   M.S. SHRIKANDE & S.S. SANE
165   Groups, combinatorics & geometry,   M.W. LIEBECK & J. SAXL (eds)
166   Surveys in combinatorics, 1991,   A.D. KEEDWELL (ed)
167   Stochastic analysis,   M.T. BARLOW & N.H. BINGHAM (eds)
168   Representations of algebras,   H. TACHIKAWA & S. BRENNER (eds)
169   Boolean function complexity,   M.S. PATERSON (ed)
170   Manifolds with singularities and the Adams-Novikov spectral sequence, B. BOTVINNIK
171   Squares,   A.R. RAJWADE
172   Algebraic varieties,   GEORGE R. KEMPF
173   Discrete groups and geometry,   W.J. HARVEY & C. MACLACHLAN (eds)
174   Lectures on mechanics,   J.E. MARSDEN
175   Adams memorial symposium on algebraic topology 1,   N. RAY & G. WALKER (eds)
176   Adams memorial symposium on algebraic topology 2,   N. RAY & G. WALKER (eds)
177   Applications of categories in computer science,   M. FOURMAN, P. JOHNSTONE, & A.. PITTS (eds)
178   Lower K- and L-theory,   A. RANICKI
179   Complex projective geometry,   G. ELLINGSRUD *et al*
180   Lectures on ergodic theory and Pesin theory on compact manifolds,   M. POLLICOTT
181   Geometric group theory I,   G.A. NIBLO & M.A. ROLLER (eds)
182   Geometric group theory II,   G.A. NIBLO & M.A. ROLLER (eds)
183   Shintani zeta functions,   A. YUKIE
184   Arithmetical functions,   W. SCHWARZ & J. SPILKER
185   Representations of solvable groups,   O. MANZ & T.R. WOLF
186   Complexity: knots, colourings and counting,   D.J.A. WELSH
187   Surveys in combinatorics, 1993,   K. WALKER (ed)
189   Locally presentable and accessible categories,   J. ADAMEK & J. ROSICKY
190   Polynomial invariants of finite groups,   D.J. BENSON
191   Finite geometry and combinatorics,   F. DE CLERCK *et al*
192   Symplectic geometry,   D. SALAMON (ed)
193   Computer algebra and differential equations,   E. TOURNIER (ed)
195   Arithmetic of blowup algebras,   WOLMER VASCONCELOS
196   Microlocal analysis for differential operators,   A. GRIGIS & J. SJÖSTRAND
197   Two-dimensional homotopy and combinatorial group theory,   C. HOG-ANGELONI,
        W. METZLER & A.J. SIERADSKI (eds)
198   The algebraic characterization of geometric 4-manifolds,   J.A. HILLMAN
199   Invariant potential theory in the unit ball of $\mathbb{C}^n$,   MANFRED STOLL

London Mathematical Society Lecture Note Series. 199

Invariant Potential Theory in the Unit Ball of $\mathbb{C}^n$

Manfred Stoll
University of South Carolina

CAMBRIDGE
UNIVERSITY PRESS

Published by the Press Syndicate of the University of Cambridge
The Pitt Building, Trumpington Street, Cambridge CB2 1RP
40 West 20th Street, New York, NY 10011-4211, USA
10 Stamford Road, Oakleigh, Melbourne 3166, Australia

First published 1994

Library of Congress cataloguing in publication data available

British Library cataloguing in publication data available

ISBN 0 521 46830 2 paperback

Transferred to digital printing 2003

Zum Andenken

an

Wladislaw Kobryn und Georg Stoll

Contents

Preface

Introduction   1

1. Notation and Preliminary Results   7

    1.1   Notation   7
    1.2   Integral Formulas on B 10
    1.3   Automorphisms of B 11

2. The Bergman Kernel   12

    2.1   The Bergman Kernel   12
    2.2   Examples   16
    2.3   Properties of the Bergman Kernel   19
    2.4   The Bergman Metric   20

3. The Laplace-Beltrami Operator   23

    3.1   The Invariant Laplacian   23
    3.2   The Invariant Laplacian for U^n 24
    3.3   The Invariant Laplacian for B 25
    3.4   The Invariant Gradient   27

4. Invariant Harmonic and Subharmonic Functions   31

    4.1 $\mathcal{M}$-Subharmonic Functions   31
    4.2   The Invariant Convolution on B 34
    4.3   The Riesz Measure   36
    4.4   Remarks   40

5. Poisson-Szegö Integrals   43

    5.1   The Poisson-Szegö Kernel   43
    5.2   The Dirichlet Problem for $\widetilde{\Delta}$ 48
    5.3   Poisson-Szegö Integrals   51
    5.4   The Dirichlet Problem for rB 55
    5.5   Remarks   57

6. The Riesz Decomposition Theorem   60

    6.1   Harmonic Majorants for $\mathcal{M}$-Subharmonic Functions   61
    6.2   The Green's Function for $\widetilde{\Delta}$ 64
    6.3   The Riesz Decomposition Theorem   67

6.4   Green Potentials   71
6.5   A Characterization of H^p Spaces   75
6.6   Remarks   79

7. Admissible Boundary Limits of Poisson Integrals   81

7.1   Admissible Limits of Poisson Integrals   82
7.2   Maximal Functions of Measures   83
7.3   Differentiation Theorems   87
7.4   The Admissible Maximal Function   89
7.5   Weighted Radial Limits of Poisson Integrals   92
7.6   Remarks   94

8. Radial and Admissible Boundary Limits of Potentials   96

8.1   Radial Limits of Potentials   96
8.2   Admissible Limits of Potentials   107
8.3   Tangential Limits of Potentials   114
8.4   Convergence in L^p 120
8.5   Related Results   123

9. Gradient Estimates and Riesz Potentials   126

9.1   Gradient Estimates of Green Potentials   126
9.2 L^p Inequalities for the Riesz Operator   132

10. Spaces of Invariant Harmonic Functions   142

10.1   Mean Value Inequalities for $|h|^p$ and $|\widetilde{\nabla}h|^p$, $0 < p < \infty$ 142
10.2   On a Theorem of Hardy and Littlewood   149
10.3 $\mathcal{M}$-Harmonic Bergman and Dirichlet Spaces   152
10.4   Remarks   160

References   164

Index   171

Preface

These notes are based on a year long seminar given by the author at the University of South Carolina during the 1992-93 academic year. The main purpose of the notes is to introduce the reader to some of the recent results in potential theory with respect to the Laplace-Beltrami operator in several complex variables, with special emphasis on the unit ball in $\mathbb{C}^n$.

The term "invariant" in the title stems from the fact that the Laplace-Beltrami operator $\widetilde{\Delta}$ of a domain is invariant under the biholomorphic mappings of the domain onto itself. Specifically, $\widetilde{\Delta}(f \circ \psi) = (\widetilde{\Delta}f) \circ \psi$ for all biholomorphic mappings ψ of the domain. Potential theory with respect to the Laplace-Beltrami operator on the ball is one of the natural extensions to several complex variables of potential theory in the unit disc. This approach is related to the non-euclidean geometry of the ball, and differs significantly from the usual euclidean extension of potential theory to the unit ball in $\mathbb{R}^n$, $n \geq 3$.

The extension to invariant potential theory in the unit ball in $\mathbb{C}^n$ for $n > 1$ does not just involve the extension of one variable techniques and results to several variables. There are significant difference between $n = 1$ and $n > 1$. Many of the classical results which are true when $n = 1$, fail to be true for the invariant Laplacian when $n > 1$. Since invariant harmonic functions are not preserved under dilations, results in the unit disc or the unit ball in $\mathbb{R}^n$, which used dilations of the domain in their arguments, require new techniques and approaches in the invariant setting. Also, since the singularities of the invariant Poisson kernel and Green's functions are non-euclidean, proofs and statements of results have a greater dependence on the non-euclidean geometry of the domain.

The study of harmonic function theory with respect to the invariant Laplacian $\widetilde{\Delta}$ gained momentum in the 1960's with the Poisson integral representation of bounded harmonic functions on symmetric spaces by Furstenberg, and the extension of the classical Fatou theorem to Poisson integrals on the ball and bounded symmetric domains by Koranyi, Stein, and Weiss. Since that time, there has been considerable research activity in the study of invariant harmonic functions, and more recently, in invariant potential theory in general. Although proofs of many of the results are available elsewhere,

the purpose of the notes is to provide a cohesive treatment of this important
subject. It is hoped that these notes will not only be useful in providing
information on the subject area, but will also stimulate additional research,
especially on other domains.

Although our primary emphasis is on potential theory with respect to the
invariant Laplacian $\widetilde{\Delta}$ on the unit ball B in $\mathbb{C}^n$, the introduction provides
an overview of the various extensions of classical potential theory to several
complex variables. Throughout the notes I have attempted to provide references to some of the known results in various other settings, and also to
additional results on the ball, the proofs of which have not been included in
the notes.

Chapters 2 and 3 deal with general results concerning the Bergman kernel, the Laplace-Beltrami operator, and the invariant gradient on bounded
domains in $\mathbb{C}^n$. In Chapters 4 through 6 we include some of the basic properties of functions harmonic and subharmonic with respect to the invariant
Laplacian in the unit ball B, including the invariant Poisson kernel, Poisson integrals, and the Riesz decomposition theorem. Chapters 7 and 8 deal
with the extension of the classical Fatou's theorem on nontangential limits of
Poisson integrals, and Littlewood's theorem on the existence of radial limits
of subharmonic functions. In Chapter 8 we also include results on admissible
boundary limits of subharmonic functions, and tangential boundary limits of
potentials. Chapter 9 contains recent results involving L^p estimates for the
invariant gradient of Green potentials, and in Chapter 10 we include several
results on weighted Bergman and Dirichlet type spaces of invariant harmonic
functions on B.

With a few exceptions, the notes are self contained, and should be accessible to anyone who has some basic knowledge of several complex variables,
measure theory, and functional analysis. Most of the references to the preliminary material in several complex variables come from the texts "Function
Theory of Several Complex Variables" by Steven Krantz, and "Function Theory in the Unit Ball of $\mathbb{C}^n$" by Walter Rudin. A standard graduate course in
real analysis should be sufficient for the prerequisites in measure theory and
functional analyis.

The author would like to thank both the faculty and graduate students
who participated in the seminar for their patience, endurance, and the many
helpful comments they provided. Special thanks go to my students K.
Adzievski and L. Rzepecki who proof-read the manuscript and corrected
many of my typing errors. Finally, the author would like to apologize to
the many researchers studying potential theory in $\mathbb{R}^n$ for neglecting to make
reference to their many contributions to the subject area.

Introduction

In the study of one complex variable, a real valued C^2 function f defined on a domain Ω in $\mathbb{C}$ is harmonic if $\Delta f = 0$, where

$$\Delta = \frac{\partial^2}{\partial x^2} + \frac{\partial^2}{\partial y^2} = 4 \frac{\partial^2}{\partial z \partial \bar{z}}$$

is the Laplacian in $\mathbb{C}$ or $\mathbb{R}^2$. In the above,

$$\frac{\partial}{\partial z} = \frac{1}{2} \left(\frac{\partial}{\partial x} - i \frac{\partial}{\partial y} \right), \qquad \frac{\partial}{\partial \bar{z}} = \frac{1}{2} \left(\frac{\partial}{\partial x} + i \frac{\partial}{\partial y} \right).$$

It is well known that f is harmonic in Ω if and only if locally f is the real part of a holomorphic function.

In addition to the usual Laplacian on Ω, there is also the invariant Laplacian or the Laplace-Beltrami operator $\tilde{\Delta}$ which is defined in terms of the Bergman kernel function of Ω. For the unit disc U, this operator is given by

$$\tilde{\Delta} = 2(1 - |z|^2)^2 \frac{\partial^2}{\partial z \partial \bar{z}}.$$

The operator $\tilde{\Delta}$ has the property that

$$\tilde{\Delta}(f \circ \psi) = (\tilde{\Delta} f) \circ \psi$$

for all automorphisms ψ of U. It is clear that $\tilde{\Delta} f = 0$ if and only if $\Delta f = 0$, and thus when $n = 1$, the euclidean and noneuclidean definitions coincide.

When one considers $\mathbb{C}^n$, $n > 1$, there are many concepts of harmonic, and as a general rule, these are all different. The extensions to $\mathbb{C}^n$ are usually separated into local definitions and global definitions. For a domain Ω in $\mathbb{C}^n$, $n > 1$, the three common local definitions are as follows:

(1) A C^2 function $f : \Omega \to \mathbb{R}$ is **harmonic** (or euclidean harmonic) if $\Delta f = 0$ in Ω, where here

$$\Delta = \sum_{j=1}^{n} \frac{\partial^2}{\partial x_j^2} + \frac{\partial^2}{\partial y_j^2} = 4 \sum_{j=1}^{n} \frac{\partial^2}{\partial z_j \partial \bar{z}_j}$$

1

is the usual Laplacian on $\mathbb{R}^{2n}$. It is well known that this is equivalent to the following: a continuous function $f : \Omega \to \mathbb{R}$ is harmonic if for every $a \in \Omega$,

$$f(a) = \int_S f(a + rt)\, d\sigma(t)$$

for all $r > 0$ sufficiently small, where S is the unit sphere in $\mathbb{C}^n$ ($\mathbb{R}^{2n}$) and σ is the normalized rotation invariant measure on S.

(2) A C^2 function $f : \Omega \to \mathbb{R}$ is **n-harmonic** if

$$\frac{\partial^2 f}{\partial z_j \partial \bar{z}_j} = 0 \qquad \text{for all} \quad j = 1, ..., n.$$

Equivalently, f is n-harmonic if f is harmonic in each variable separately.

(3) A function $f : \Omega \to \mathbb{R}$ is **pluriharmonic** if for each $a \in \Omega$, $b \in \mathbb{C}^n$, the function

$$\lambda \to f(a + \lambda b)$$

is harmonic in a neighborhood of 0 in $\mathbb{C}$. It is easily shown that a C^2 function f is pluriharmonic if and only if

$$\frac{\partial^2}{\partial z_j \partial \bar{z}_k} = 0$$

for all j, k. This is equivalent to f being locally the real part of a holomorphic function.

From the definitions it is clear that

$$(3) \ \subset \ (2) \ \subset \ (1) ,$$

and that all containments are proper.

In addition to the above, there are also global definitions which depend on the domain. If Ω is a domain in $\mathbb{C}^n$, we let $Aut(\Omega)$ denote the group of biholomrphic mappings of Ω onto itself. The domain Ω is said to be **homogeneous** if $Aut(\Omega)$ is transitive on Ω, i.e., for $z_1, z_2 \in \Omega$, there exists $\psi \in Aut(\Omega)$ such that $\psi(z_1) = z_2$. A homogeneous domain Ω is **symmetric** if for any $z_o \in \Omega$, there exists $\psi \in Aut(\Omega)$ such that

(a) $\psi(z_o) = z_o$ but $\psi(z) \neq z$ for all $z \neq z_o$, and

(b) $\psi \circ \psi = $ identity on Ω.

Both the unit ball B_n and the unit polydisc U^n are bounded symmetric domains. In $\mathbb{C}^2$, every bounded homogeneous domain is symmetric and is biholomorphic to either B_2 or U^2. In $\mathbb{C}^3$, every bounded homogeneous domain is again symmetric and is biholomorphic to one of the following:

(i) B_3,   (ii) $B_2 \times U$,   (iii) U^3,

or the future light cone

(iv) $\{(z_1, z_2, z_3) : y_3 > \sqrt{y_1^2 + y_2^2}\}$.

In $\mathbb{C}^n$, the number of nonequivalent bounded symmetric domains is finite. E. Cartan [Ca] proved that there exist only six types of irreducible bounded symmetric domains, the so called four classical domains and two exceptional domains of dimensions 16 and 27 respectively. The book by L. K. Hua ([Hu]) is an excellent introduction to harmonic analysis on the classical Cartan domains. Closely related to the bounded symmetric domains are the generalized half-planes or Siegel domains of type II ([Gi, Ko1]). Every bounded symmetric domain has a realization of this type.

Let Ω be a domain in $\mathbb{C}^n$. A differential operator D on $C^\infty(\Omega)$ is said to be invariant if

$$D(f \circ \psi) = (D f) \circ \psi \qquad \text{for all } \psi \in Aut(\Omega).$$

It is well known ([He1, He2]) that if Ω is a symmetric domain, then the algebra of invariant differential operators on $C^\infty(\Omega)$ is finitely generated. It is also known that the Laplace-Beltrami operator $\widetilde{\Delta}_\Omega$ with respect to the Bergman metric on a bounded domain Ω is invariant.

The two common global definitions of harmonic functions on a bounded domain in $\mathbb{C}^n$ are as follows:

(4) A C^2 function $f : \Omega \to \mathbb{R}$ is **invariant harmonic** or **weakly harmonic** if $\widetilde{\Delta}_\Omega f = 0$, where $\widetilde{\Delta}_\Omega$ is the Laplace-Beltrami operator on Ω.

(5) A C^∞ function $f : \Omega \to \mathbb{R}$ is **strongly harmonic** if $D f = 0$ for all invariant differential operators D on $C^\infty(\Omega)$ for which $D 1 = 0$.

Since the Laplace-Beltrami operator is invariant, we clearly have (5) $\subset$ (4), and except in the case of the ball, the containment is proper. For symmetric domains, or symmetric spaces in general, there is an abundance of information about strongly harmonic functions. A nice survey of the subject matter may be found in [Ko3, Ko4]. In 1963, Furstenberg [Fu1] (see also [Fu2]) using probability theory and Lie group machinery obtained a Poisson integral representation for bounded strongly harmonic functions on Riemannian symmetric spaces of noncompact type, which includes the bounded symmetric domains. In the same paper he also proved that in this setting every bounded weakly harmonic function is strongly harmonic.

For arbitrary domains, there are very few results concerning weakly harmonic and subharmonic functions. In the notes we will develop many of the basic properties of invariant (or weakly) harmonic and subharmonic functions on the unit ball of $\mathbb{C}^n$. In the remarks we will point out some of the results which are known in other settings such as the polydisc, or bounded symmetric domains. In many instances however, the solutions of the analogous problems, even for simple domains such as the polydisc, or the pseudoconvex ellipsoidal domains

$$D_\alpha = \{(z_1, z_2) : |z_1|^{2/\alpha} + |z_2|^2 < 1\}, \quad \alpha \neq 1,$$

are in most cases unknown.

For the polydisc U^n, there is a significant difference between strongly harmonic and weakly harmonic functions. As we will see in Sections 3.2 and 4.4, a function on U^n is strongly harmonic (subharmonic) if and only if it is n-harmonic (n-subharmonic). In Section 3.2, we show that for U^n, the Laplace-Beltrami operator $\widetilde{\Delta}_{U^n}$ is given by

$$\widetilde{\Delta}_{U^n} = 2 \sum_{j=1}^{n} (1 - |z_j|^2)^2 \frac{\partial^2}{\partial z_j \partial \overline{z}_j},$$

and we also give an example of a function f which is weakly harmonic, but not strongly harmonic. Even though strongly harmonic and subharmonic functions on the polydisc have been studied since the late 1930's, the few results concerning weakly harmonic functions are much more recent. Fatou's theorem for weakly harmonic functions on U^n was not proved until 1983 (see Section 7.6), and at present, the appropriate analogue of Littlewood's theorem for weakly subharmonic functions appears to be unknown.

As was indicated above, for the unit ball B, the concepts of weakly harmonic and strongly harmonic functions coincide. In Section 3.3, we will show that the Laplace-Beltrami operator on B is given by

$$\widetilde{\Delta}_B = \frac{4}{(n+1)}(1 - |z|^2) \sum_{i,j=1}^{n} [\delta_{i,j} - \overline{z}_i z_j] \frac{\partial^2}{\partial z_j \partial \overline{z}_i}.$$

Functions on B which are harmonic with respect to this operator are ususaly referred to as **invariant** harmonic or $\mathcal{M}$-**harmonic**. As we will see in Section 4.1, this is equivalent to f satisfying the non-euclidean mean value property

$$f(\psi(0)) = \int_S f(\psi(rt)) \, d\sigma(t), \qquad 0 < r < 1,$$

for all $\psi \in Aut(B)$.

There are some significant differences between the usual Laplacian Δ as given in (1), and the Laplace-Beltrami operator $\widetilde{\Delta}_B$. The ordinary Laplacian Δ is uniformly elliptic on B. This is not the case for the operator $\widetilde{\Delta}$, which is degenerate on the boundary. Since Δ is uniformly elliptic on B, if f is C^∞ on the boundary S of B, and if F is the solution to the Dirichlet problem for Δ, then F is C^∞ on $\overline{B}$. For the operator $\widetilde{\Delta}$ this result fails dramatically. In

Section 5.2 we give an example of a C^∞ function on S for which the solution F of the Dirichlet problem for $\widetilde{\Delta}$ is not C^2 on S.

Another significant difference is as follows: for the usual Laplacian, $\Delta f_r = r^2 \Delta f$, where for $0 < r < 1$, $f_r(z) = f(rz)$. Thus if f is harmonic or subharmonic, the same is true for f_r. This fails for the operator $\widetilde{\Delta}$, and as a consequence, many of the classical results about harmonic or subharmonic functions which relied on contractions, require new proofs. Even standard results, such as proving that a bounded harmonic function can be represented as the Poisson integral of a bounded function, become nontrivial for $\mathcal{M}$-harmonic functions on the ball. It is partly for this reason that the Poisson integral formula for bounded strongly harmonic functions remained unsolved until the work of Furstenberg in 1963.

In Sections 5.1 and 5.2 we introduce the Poisson-Szegö kernel and consider the Dirichlet problem for $\widetilde{\Delta}$ on the ball. These results seem to have appeared first in 1958 in the book by L. K. Hua [Hu]. In Section 5.3 we prove the Poisson integral formula for functions in the Hardy space $\mathcal{H}^p$, $1 \le p \le \infty$, of $\mathcal{M}$-harmonic functions on B. The Poisson integral formula itself follows from a more general result about $\mathcal{M}$-subharmonic functions. In Theorem 5.8, using an equicontinuity argument due to Ullrich, we prove that if f is a continuous $\mathcal{M}$-subharmonic function on B satisfying

$$\sup_{0 < r < 1} \int_S |f(rt)|^p d\sigma(t) < \infty, \qquad 1 \le p < \infty,$$

or is bounded when $p = \infty$, then f has an $\mathcal{M}$-harmonic majorant which is the Poisson-Szegö integral of an L^p function when $p > 1$, and a measure when $p = 1$. The argument given also proves the Poisson integral formula for functions in the Hardy space $\mathcal{H}^p$, $1 \le p \le \infty$.

In Chapter 6 we derive the Green's function for $\widetilde{\Delta}$ and prove the Riesz decomposition theorem for $\mathcal{M}$-subharmonic functions having an $\mathcal{M}$-harmonic majorant. As an application of the results of this chapter, we prove that a harmonic function f is in $\mathcal{H}^p$, $1 < p < \infty$, if and only if

$$\int_B (1 - |z|^2)^n |f(z)|^{p-2} |\widetilde{\nabla} f(z)|^2 d\lambda(z),$$

where $\widetilde{\nabla}$ denotes the invariant gradient on B, and λ is the invariant volume measure on B. The above also characterizes the classical Hardy spaces H^p of holomorphic functions on B for all p, $0 < p < \infty$. When $n = 1$, $|\widetilde{\nabla} f(z)|^2 = 2(1 - |z|^2)^2 |f'(z)|^2$, for holomorphic functions f.

Chapters 7 and 8 are devoted to extensions to the ball of Fatou's theorem and Littlewood's theorem concerning boundary limits of $\mathcal{M}$-harmonic and

$\mathcal{M}$-subharmonic functions on B. For Fatou's theorem, the significant difference between the euclidean and non-euclidean case is that the appropriate
analogue of the non-tangential approach regions are the admissible regions
of Koranyi. For a point $\zeta \in S$, these are non-tangential along the complex
line $\{\lambda\zeta : \lambda \in \mathbb{C}, |\lambda| < 1\}$, but are tangential in the orthogonal direction.

In Chapter 8, in addition to proving Littlewood's theorem concerning
radial limits of $\mathcal{M}$-subharmonic functions, we also prove the existence of
admissible limits for $\mathcal{M}$-subharmonic functions f for which the Riesz measure
$\widetilde{\Delta}f$ is absolutely continuous and satisfies

$$\int_B (1 - |w|^2)^n (\widetilde{\Delta}f(w))^p \, d\lambda(w) < \infty,$$

for some $p > n$. Section 8.3 also contains recent results on tangential limits
of invariant Green potentials in B. In Chapter 9 we present several results
concerning invariant Riesz potentials and L^p estimates for the invariant gradient of Green potentials of functions on B. These are motivated by well
known results about the classical Riesz kernel on $\mathbb{R}^n$.

In the final chapter we include some recent results on weighted Bergman
and Dirichlet type spaces of $\mathcal{M}$-harmonic functions on B. Section 10.1 con-
tains several useful mean value inequalities for both $|h|^p$ and $|\widetilde{\nabla}h|^p$, $0 < p <$
∞, for $\mathcal{M}$-harmonic functions h on B. These inequalities are used in the
final two sections to first prove a generalization of a theorem of Hardy and
Littlewood comparing the rate of growth of the p'th means of h and $\widetilde{\nabla}h$,
and in the study of weighted Bergman and Dirichlet type spaces on B. For
$0 < p < \infty$, $\gamma \in \mathbb{R}$, the weighted Dirichlet space $\mathcal{D}_p^\gamma$ is defined as the space
of $\mathcal{M}$-harmonic functions h on B for which

$$\int_B (1 - |z|^2)^\gamma |\widetilde{\nabla}h(z)|^p \, d\lambda(z) < \infty.$$

Similarly, the weighted Bergman space $\mathcal{A}_p^\gamma$ consists of those $\mathcal{M}$-harmonic
functions h for which

$$\int_B (1 - |z|^2)^\gamma |h(z)|^p \, d\lambda(z) < \infty.$$

One of the main results of Section 10.3 is that for $p \geq 1$, $\gamma > n$, $h \in \mathcal{A}_p^\gamma$ if
and only if $h \in \mathcal{D}_p^\gamma$.

1.
Notation and Preliminary Results

In this chapter we introduce some basic notation and some preliminary results which will be used throughout the notes. The notation is as in [Ru3], and many of the proofs of some of the preliminary results can be found in that text.

1.1. Notation.

As usual, $\mathbb{C}^n$ will denote n-dimensional complex space with inner product

$$\langle z, w \rangle = \sum_{j=1}^{n} z_j \overline{w}_j, \qquad (z, w \in \mathbb{C}^n),$$

and the associated norm $|z| = \sqrt{\langle z, z \rangle}$. The standard orthonormal basis elements in $\mathbb{C}^n$ are denoted by $e_1, ..., e_n$. For $a \in \mathbb{C}^n$, $r > 0$, let

$$B(a, r) = \{z \in \mathbb{C}^n : |z - a| < r\}.$$

For simplicity, the unit ball $B(0, 1)$ will be denoted by either B or B_n. The boundary of B is the unit sphere $S = \{z : |z| = 1\}$. When $n = 1$, the unit disc in $\mathbb{C}$ will be denoted by U, and for $n > 1$,

$$U^n = \{z \in \mathbb{C}^n : \Delta(z) < 1\},$$

where

$$\Delta(z) = \max\{|z_j| : j = 1, 2, ..., n\}.$$

The set $T^n = \{z : |z_j| = 1, j = 1, 2, ..., n\}$ is called the distinguished boundary of U^n.

Since our discussion of functions of several complex variables requires multi-index notation, we introduce the following standard conventions. Let $N = \{0, 1, 2, ...\}$ denote the set of natural numbers. A multi-index α is an ordered n-tuple

$$\alpha = (\alpha_1, ..., \alpha_n) \qquad \text{with } \alpha_j \in N, j = 1, ..., n.$$

For a multi-index α and $z \in \mathbb{C}^n$, set

$$|\alpha| = \alpha_1 + \cdots + \alpha_n,$$
$$\alpha! = \alpha_1! \cdots \alpha_n!,$$
$$z^\alpha = z_1^{\alpha_1} \cdots z_n^{\alpha_n}.$$

If Ω is an open subset of $\mathbb{C}^n$, $k \in N$, we denote by $C^k(\Omega)$ the space of real or complex valued function on Ω which have continuous derivatives of order α for all multi-indices α with $|\alpha| \leq k$. $C_c^k(\Omega)$ denotes those functions in $C^k(\Omega)$ with compact support. The meanings of $C^\infty(\Omega)$ and $C_c^\infty(\Omega)$ are as usual.

Let Ω be an open subset of $\mathbb{C}^n$. A function $f : \Omega \to \mathbb{C}$ is **holomorphic** if f is holomorphic in each variable separately, i.e., for each $a \in \Omega$ and each $i = 1, ..., n$, the function

$$\lambda \to f(a + \lambda e_i)$$

is holomorphic in an open neighborhood of 0 in $\mathbb{C}$. It is an old result due to Hartogs that if f is holomorphic as defined above, then f is continuous in Ω, and as a consequence of the Cauchy integral formula for polydiscs, for every $a \in \Omega$, f has a power series expansion

$$f(z) = \sum_\alpha a_\alpha (z - a)^\alpha,$$

which converges absolutely in a neighborhood $B(a, r)$ of a, and uniformly on compact subsets of $B(a, r)$, where for $\alpha = (\alpha_1, ..., \alpha_n)$,

$$a_\alpha = \frac{\partial^{|\alpha|} f}{\partial z_1^{\alpha_1} \cdots \partial z_n^{\alpha_n}}(a).$$

As in the case $n = 1$, if $z_j = x_j + i y_j$, with x_j and y_j real, the partial differential operators $\frac{\partial}{\partial z_j}$ and $\frac{\partial}{\partial \bar{z}_j}$ are defined by

$$\frac{\partial}{\partial z_j} = \frac{1}{2}\left(\frac{\partial}{\partial x_j} - i \frac{\partial}{\partial y_j} \right), \quad \text{and} \quad \frac{\partial}{\partial \bar{z}_j} = \frac{1}{2}\left(\frac{\partial}{\partial x_j} + i \frac{\partial}{\partial y_j} \right).$$

If f is holomorphic, it is an immediate consequence of the definition that

$$\frac{\partial f}{\partial \bar{z}_j} = 0 \quad \text{for all} \quad j = 1, ..., n.$$

For a C^1 function f, this condition is also sufficient that f be holomorphic.

If $w = \varphi(z) = (\varphi_1(z), ..., \varphi_k(z))$ is a C^1 mapping of a domain $\Omega_1 \subset \mathbb{C}^n$ into a domain $\Omega_2 \subset \mathbb{C}^k$, and f is a C^1 function on Ω_2, then for $g(z) = f(\varphi(z))$, the following complex versions of the chain rule hold:

$$(1.1) \qquad \frac{\partial g}{\partial z_j} = \sum_{i=1}^{k} \left(\frac{\partial f}{\partial w_i} \frac{\partial w_i}{\partial z_j} + \frac{\partial f}{\partial \overline{w}_i} \frac{\partial \overline{w}_i}{\partial z_j} \right)$$

and

$$(1.2) \qquad \frac{\partial g}{\partial \overline{z}_j} = \sum_{i=1}^{k} \left(\frac{\partial f}{\partial w_i} \frac{\partial w_i}{\partial \overline{z}_j} + \frac{\partial f}{\partial \overline{w}_i} \frac{\partial \overline{w}_i}{\partial \overline{z}_j} \right)$$

If φ is a holomorphic mapping, i.e., $\varphi_j(z)$ is holomorphic for all $j = 1, ..., k$, then $\dfrac{\partial w_i}{\partial \overline{z}_j} = 0$ for all i, j. Thus if f is holomorphic, $f \circ \varphi$ is also holomorphic.

If Ω_1, Ω_2 are domains in $\mathbb{C}^n$, a one-to-one holomorphic mapping φ of Ω_1 onto Ω_2 with holomorphic inverse is called a **biholomorphic mapping** of Ω_1 onto Ω_2. That φ is one-to-one and onto is sufficient that the inverse map is holomorphic and that $\det J_\varphi(z) \neq 0$ for all $z \in \Omega_1$, where J_φ is the Jacobian matrix of the mapping φ given by

$$(1.3) \qquad J_\varphi(z) = \left(\frac{\partial \varphi_i(z)}{\partial z_j} \right)_{i,j=1}^{n} .$$

For a domain $\Omega \subset \mathbb{C}^n$, the group (under composition) of all biholomorphic mappings of Ω onto Ω is called the automorphism group of Ω and is denoted by $Aut(\Omega)$.

Let Ω be a domain in $\mathbb{C}^n$. An upper semicontinuous function $f : \Omega \to [-\infty, \infty)$ with $f \not\equiv -\infty$ is **plurisubharmonic** (p.s.h.) on Ω if for each $a \in \Omega$, $w \in \mathbb{C}^n$, the function

$$(1.4) \qquad \lambda \to f(a + \lambda w)$$

is subharmonic in a neighborhood of 0 in $\mathbb{C}$. A continuous function $f : \Omega \to \mathbb{R}$ is **pluriharmonic** if the function (1.4) is harmonic in a neighborhood of 0 in $\mathbb{C}$ for every $a \in \Omega$ and $w \in \mathbb{C}^n$.

For a C^2 function f, if one computes the Laplacian of the function (1.4) at 0 by using the complex form of the chain rules (1.1) and (1.2), one obtains that f is p.s.h. on Ω if and only if

$$(1.5) \qquad \sum_{i,j=1}^{n} \frac{\partial^2 f(z)}{\partial z_i \partial \overline{z}_j} w_i \overline{w}_j \geq 0$$

for all $z \in \Omega$ and $w \in \mathbb{C}^n$.

1.2. Integral Formulas on B.

Let ν denote Lebesgue measure in $\mathbb{C}^n$ normalized so that $\nu(B) = 1$. If V denotes the usual Euclidean volume measure in $\mathbb{C}^n$ and $c_n = V(B_n)$, then $c_n d\nu = dV$. Also, let σ denote the rotation-invariant measure on S also normalized so that $\sigma(S) = 1$. The term rotation-invariant refers to the orthogonal group $O(2n)$ of all isometries of $\mathbb{R}^{2n}$ which fix the origin.

The following integration formulas, the proofs of which may be found in [Ru3], will prove useful throughout.

$$(1.6) \qquad \int_{\mathbb{C}^n} f \, d\nu = 2n \int_0^\infty r^{2n-1} \int_S f(r\zeta) \, d\sigma(\zeta) \, dr.$$

$$(1.7) \qquad \int_S f \, d\sigma = \int_S \frac{1}{2\pi} \int_0^{2\pi} f(e^{i\theta}\zeta) \, d\theta \, d\sigma(\zeta).$$

$$(1.8) \qquad \int_S f \, d\sigma = \int_{B_{n-1}} \frac{1}{2\pi} \int_0^{2\pi} f(\zeta', e^{i\theta}\zeta_n) \, d\theta \, d\nu(\zeta').$$

$$(1.9) \qquad \int_S f \, d\sigma = \int_{\mathcal{U}} f(U\eta) \, dU.$$

In (1.8), $\zeta' = (\zeta_1, ..., \zeta_{n-1})$, and in (1.9), dU denotes the Haar measure on the group $\mathcal{U} = \mathcal{U}(n)$ of unitary transformations of $\mathbb{C}^n$. A linear transformation $U \in \mathcal{U}$ if and only if

$$\langle Uz, Uw \rangle = \langle z, w \rangle$$

for all $z, w \in \mathbb{C}^n$. The group $\mathcal{U}$ is a compact subgroup of $O(2n)$. In addition to the above, if f is a function of one complex variable only, then for $n > 1$, $\eta \in S$,

$$(1.10) \qquad \int_S f(\langle \zeta, \eta \rangle) \, d\sigma(\zeta) = \frac{n-1}{\pi} \iint_U (1 - r^2)^{n-2} f(re^{i\theta}) \, r \, dr \, d\theta.$$

Since we will have occassion to use both the binomial and multinomial expansion formulas, we include them at this point. For $|\lambda| < 1$, β not a negative integer,

$$(1.11) \qquad (1 - \lambda)^{-\beta} = \sum_{k=0}^\infty \frac{\Gamma(k + \beta)}{\Gamma(\beta) \, k!} \lambda^k,$$

where Γ denotes the Gamma function. Also, for $k = 1, 2, ...,$

$$(1.12) \qquad \langle z, w \rangle^k = \sum_{|\alpha|=k} \frac{k!}{\alpha!} z^\alpha \overline{w}^\alpha.$$

1.3. Automorphisms of B.

Since our primary emphasis is on potential theory in the ball, we include at this point some of the basic properties of the automorphism group of B. In the case of the unit disc U, for $a \in U$, we let φ_a denote the biholomorphic mapping of U given by

$$\varphi_a(z) = \frac{a - z}{1 - \bar{a}z}.$$

These mappings are known as the Möbius transformations of U. It is easily shown that the mapping φ_a satisfies $\varphi_a(0) = a$, $\varphi_a(a) = 0$, and $\varphi_a^{-1} = \varphi_a$. The automorphism group of U is generated by the mappings $\{\varphi_a : a \in U\}$ along with the rotations.

The extension to the unit ball B is done similary. Fix $a \in B$, and let P_a be the orthogonal projection of $\mathbb{C}^n$ onto the subspace generated by a, which is explicitly given by $P_0 = 0$, and

$$P_a z = \frac{\langle z, a \rangle}{\langle a, a \rangle} a, \qquad \text{if } a \neq 0.$$

Let $Q_a = I - P_a$. Define φ_a on B by

$$(1.13) \qquad \varphi_a(z) = \frac{a - P_a z - \sqrt{1 - |a|^2}\, Q_a z}{1 - \langle z, a \rangle}.$$

If $a = (a_1, 0, ..., 0) = a_1 e_1$ with $|a_1| < 1$, this mapping is of the form

$$(1.14) \qquad \varphi_{a_1 e_1}(z) = \left(\frac{a_1 - z_1}{1 - \bar{a}_1 z_1}, -\frac{\sqrt{1 - |a_1|^2}\, z'}{1 - \bar{a}_1 z_1} \right)$$

where $z' = (z_2, ..., z_n)$.

It is easily shown that the mapping $\varphi_a \in Aut(B)$ and satisfies,

$$\varphi_a(0) = a, \quad \varphi_a(a) = 0, \quad \varphi_a(\varphi_a(z)) = z.$$

Furthermore, for all $z, w \in \overline{B}$,

$$(1.15) \qquad 1 - \langle \varphi_a(z), \varphi_a(w) \rangle = \frac{(1 - \langle a, a \rangle)(1 - \langle z, w \rangle)}{(1 - \langle z, a \rangle)(1 - \langle a, w \rangle)}.$$

In particular, for $a \in B$, $z \in \overline{B}$,

$$(1.16) \qquad 1 - |\varphi_a(z)|^2 = \frac{(1 - |a|^2)(1 - |z|^2)}{|1 - \langle z, a \rangle|^2}.$$

Finally, if $\psi \in Aut(B)$ and $a = \psi^{-1}(0)$, then there exists a unique $U \in \mathcal{U}(n)$ such that $\psi = U\varphi_a$. This result follows from the fact that $\psi \circ \varphi_a$ is an automorphism of B which fixes 0, and thus by Cartan's theorem is linear, and hence must be unitary. The group $Aut(B)$ is sometimes also called the Möbius group of B and is denoted by $\mathcal{M}$.

2.
The Bergman Kernel

In this chapter we introduce the Bergman kernel for a bounded domain in $\mathbb{C}^n$, and prove some preliminary results about the Bergman kernel and the Bergman metric which will be needed in subsequent sections.

2.1. The Bergman Kernel.

Let Ω be a bounded domain in $\mathbb{C}^n$ and let

$$A^2(\Omega) = \left\{ f \text{ holomorphic on } \Omega : \int_\Omega |f|^2 \, dV < \infty \right\}.$$

For $f, g \in A^2(\Omega)$, define

$$\langle f, g \rangle = \int_\Omega f \, \bar{g} \, dV.$$

With this inner product $A^2(\Omega)$ is an inner product space, and as a consequence of the following lemma is complete in the norm $\|f\|_{A^2} = \sqrt{\langle f, f \rangle}$.

Lemma 2.1. *If K is a compact subset of Ω, then there exists a constant C_K such that*

$$(2.1) \qquad |f(z)| \leq C_K \|f\|_{A^2} \qquad \text{for all } z \in K.$$

Proof. Since K is a compact subset of Ω, there exists an $r > 0$ such that $B(z, r) \subset \Omega$ for all $z \in K$. Let $z \in K$ and $\zeta \in S$. Then the function $\lambda \to f(z + \lambda\zeta)$ is holomorphic for all $\lambda \in \mathbb{C}$ with $|\lambda| < r$. Thus $|f(z + \lambda\zeta)|^2$ is a subharmonic function in λ for $|\lambda| < r$, and hence

$$|f(z)|^2 \leq \frac{1}{2\pi} \int_0^{2\pi} |f(z + \rho e^{i\theta}\zeta)|^2 \, d\theta,$$

for all ρ, $0 < \rho < r$. Integrating over S and using (1.7) gives

$$|f(z)|^2 \leq \int_S |f(z + \rho\zeta)|^2 \, d\sigma(\zeta).$$

Multiplying both sides by ρ^{2n-1} and integrating from 0 to r gives

$$|f(z)|^2 \le \frac{2n}{r^{2n}} \int_0^r \rho^{2n-1} \int_S |f(z+\rho\zeta)|^2 \, d\sigma(\zeta) \, d\rho$$

(2.2)
$$= \frac{1}{r^{2n}} \int_{B(z,r)} |f|^2 \, d\nu$$

$$\le \frac{1}{V(B)r^{2n}} \int_\Omega |f|^2 \, dV,$$

which proves the result. $\square$

As a consequence of Lemma 2.1, convergence in norm implies uniform convergence on compact subsets of Ω, from which it then follows that $A^2(\Omega)$ is complete.

Fix a point $z \in \Omega$. By the lemma, the functional e_z given by

$$e_z(f) = f(z), \qquad f \in A^2(\Omega),$$

is continuous. Thus by the Riesz representation theorem there exists $k_z \in A^2(\Omega)$ such that

(2.3)
$$f(z) = \langle f, k_z \rangle = \int_\Omega f(w)\overline{k_z(w)} \, dV(w).$$

Definition. *The function*

$$K(z,w) = \overline{k_z(w)}$$

is called the **Bergman kernel** *of Ω.*

Equation (2.3) is the reproducing property of the Bergman kernel, and can be rewritten as

(2.3a)
$$f(z) = \int_\Omega f(w)K(z,w) \, dV(w)$$

Since $k_z \in A^2$, we have

$$k_z(w) = \overline{K(z,w)} = \int_\Omega \overline{K(z,\zeta)}K(w,\zeta) \, dV(\zeta)$$

$$= \overline{\int_\Omega \overline{K(w,\zeta)}K(z,\zeta) \, dV(\zeta)} = K(w,z).$$

Thus $K(z,w) = \overline{K(w,z)}$, and as a consequence, for fixed $w \in \Omega$, $z \to K(z,w)$ is in $A^2(\Omega)$. Furthermore, by the above

(2.4)
$$K(z,z) = \int_\Omega |K(z,\zeta)|^2 \, dV(\zeta),$$

and thus by the Cauchy-Schwarz inequality,

$$(2.5) \qquad |K(z,w)|^2 \leq K(z,z)\,K(w,w).$$

From (2.4) it follows that $K(z,z) \geq 0$. However, if $A^2(\Omega)$ contains enough functions to separate points of Ω (which it does if Ω is bounded), then $K(z,z) > 0$ for all $z \in \Omega$. Finally, it is easy to see that the Bergman kernel is unique, i.e., if $H(z,\zeta)$ is any function satisfying $H(\cdot,\zeta) \in A^2(\Omega)$ for which the reproducing property (2.3) holds, then $H = K$.

Since $A^2(\Omega)$ is separable, it has a countable complete orthonormal system $\{\psi_k(z)\}$. Thus every $f \in A^2$ has a Fourier series expansion

$$f = \sum_{k=0}^{\infty} a_k\,\psi_k, \qquad a_k = \langle f, \psi_k \rangle,$$

where the convergence is in the norm of A^2. But as a consequence of (2.1), for all $z \in \Omega$,

$$(2.6) \qquad f(z) = \sum_{k=0}^{\infty} a_k\,\psi_k(z),$$

where the series converges absolutely in Ω, and uniformly on compact subsets of Ω. For the Bergman kernel, we also have the following:

Proposition 2.2. *Let $\{\psi_k(z)\}$ be a countable complete orthonormal system for $A^2(\Omega)$. Then*

$$(2.7) \qquad K(z,w) = \sum_{k=0}^{\infty} \psi_k(z)\overline{\psi_k(w)},$$

where the series converges absolutely, and uniformly on compact subsets of $\Omega \times \Omega$.

Proof. Since $2|\psi_k(z)\overline{\psi_k(w)}| \leq |\psi_k(z)|^2 + |\psi_k(w)|^2$, it suffices to show that the series $\sum |\psi_k(z)|^2$ converges uniformly to $K(z,z)$ on compact subsets of Ω.

Let K be a fixed compact subset of Ω, and choose $r > 0$ such that $\overline{B(a,r)} \subset \Omega$ for all $a \in K$. By the orthogonality of the $\{\psi_k\}$ and inequality (2.2),

$$\sum_{k=0}^{m} |\psi_k(a)|^2 = \int_{\Omega} \left| \sum_{k=0}^{m} \psi_k(z)\overline{\psi_k(a)} \right|^2 dV(z)$$

$$\geq \int_{B(a,r)} \left| \sum_{k=0}^{m} \psi_k(z)\overline{\psi_k(a)} \right|^2 dV(z)$$

$$\geq V(B)r^{2n} \left| \sum_{k=0}^{m} |\psi_k(a)|^2 \right|^2.$$

The last inequality follows from the fact that $z \to |\sum_0^m \psi_k(z)\overline{\psi_k(a)}|^2$ is p.s.h. in Ω. Thus

$$\sum_{k=0}^{\infty} |\psi_k(a)|^2 \leq \frac{1}{V(B)r^{2n}}$$

for all $a \in K$. Consequently $K(z,z)$ is bounded on every compact subset of Ω. Let

$$K_r = \bigcup_{a \in K} \overline{B(a,r)}.$$

Then K_r is a compact subset of Ω and by Lemma 2.1, there exists a constant C_r such that

$$|f(z)|^2 \leq C_r \int_{K_r} |f(\zeta)|^2 \, dV(\zeta)$$

for all $z \in K$. Since

$$\sum_{k=0}^{\infty} \int_{K_r} |\psi_k(\zeta)|^2 \, dV(\zeta) = \int_{K_r} K(\zeta,\zeta) \, dV(\zeta) < \infty,$$

given $\epsilon > 0$, there exists an integer M such that

$$\sum_{M+1}^{\infty} |\psi_k(z)|^2 \leq C_r \sum_{M+1}^{\infty} \int_{K_r} |\psi_k(\zeta)|^2 \, dV(\zeta) < \epsilon$$

for all $z \in K$. Therefore $\sum_{k=0}^{\infty} |\psi_k(z)|^2$ converges uniformly to $K(z,z)$ on K. $\square$

We now show how $K(z,w)$ can be computed for a wide class of domains. A subset R of $\mathbb{C}^n$ is called a **Reinhardt set** if $z \in R$ implies $(e^{i\theta_1}z_1, ..., e^{i\theta_n}z_n) \in R$ for all $0 \leq \theta_j < 2\pi$, $j = 1, ..., n$.

Proposition 2.3. *Let Ω be a bounded Reinhardt domain with $0 \in \Omega$. Then there exists a complete orthonormal system $\{\psi_{k,\nu}(z)\}$,*

$$\nu = 1, ..., m_k = \binom{n+k-1}{k}, \qquad k = 0, 1, 2, ...$$

of homogeneous polynomials of degree k.

Proof. For a Reinhardt domain Ω with $0 \in \Omega$, it is well known ([Kr1, Proposition 2.3.14]) that if f is holomorphic on Ω, then

$$(2.8) \qquad f(z) = \sum_{k=0}^{\infty} \sum_{|\alpha|=k} a_\alpha z^\alpha,$$

where the series converges absolutely in Ω, and uniformly on compact subsets of Ω.

For each $k = 0, 1, 2, ...$, there are exactly m_k monomials z^α with $|\alpha| = k$. Denote these by $Z_{k,\nu}$, and let $\psi_{k,\nu} = b_{k,\nu} Z_{k,\nu}$, where the $b_{k,\nu}$ are chosen so that

$$\int_\Omega |\psi_{k,\nu}|^2 \, dV = 1.$$

Consider $\langle \psi_{k,\nu}, \psi_{j,\alpha} \rangle$. Since dV is invariant under coordinatewise multiplication by $e^{i\theta_j}$,

$$\int_\Omega f(z) \, dV(z) = \int_\Omega \frac{1}{(2\pi)^n} \int_0^{2\pi} \cdots \int_0^{2\pi} f(e^{i\theta_1} z_1, ..., e^{i\theta_n} z_n) \, d\theta_1 \cdots d\theta_n \, dV(z),$$

and as a consequence

$$\langle \psi_{k,\nu}, \psi_{j,\alpha} \rangle = 0 \quad \text{if } k \neq j \text{ or } \nu \neq \alpha \text{ when } k = j.$$

Thus $\{\psi_{k,\nu}(z)\}$ is orthonormal.

Suppose K is a compact subset of Ω which is Reinhardt. Then by (2.8) and orthogonality,

$$\int_K f(z)\overline{\psi_{k,\nu}(z)} \, dV(z) = a_{k,\nu} \int_K Z_{k,\nu}(z)\overline{\psi_{k,\nu}(z)} \, dV(z)$$

$$= \frac{a_{k,\nu}}{b_{k,\nu}} \int_K |\psi_{k,\nu}|^2 \, dV.$$

Since $\Omega = \cup_{j=1}^\infty K_j$, where each K_j is compact and Reinhardt with $K_j \subset K_{j+1}$, we obtain

$$a_{k,\nu} = b_{k,\nu} \langle f, \psi_{k,\nu} \rangle.$$

Such a sequence $\{K_j\}$ can be obtained by letting

$$K_j = \{z \in \Omega : \Delta_\Omega(z) \geq 1/j\},$$

where $\Delta_\Omega(z) = \inf\{\Delta(z - \zeta) : \zeta \in \Omega^c\}$. Thus by (2.8), if $f \in A^2(\Omega)$,

$$f(z) = \sum_{k=0}^\infty \sum_{\nu=1}^{m_k} \langle f, \psi_{k,\nu} \rangle \psi_{k,\nu}(z),$$

where the series converges absolutely, and uniformly on compact subsets of Ω. As a consequence, $\{\psi_{k,\nu}\}$ is a complete orthonormal system for $A^2(\Omega)$. $\square$

2.2. Examples.

(1) Let $\Omega = B_n$. By the previous theorem, a complete othonormal system is given by $\{b_\alpha z^\alpha\}$ where the b_α are such that

$$\int_B |b_\alpha z^\alpha|^2 \, dV(z) = |b_\alpha|^2 \int_B |z^\alpha|^2 \, dV(z) = 1.$$

To determine the b_α we need the following lemma:

Lemma 2.4. *For every multi-index α,*

$$(2.9) \qquad \int_S |\zeta^\alpha|^2 \, d\sigma(\zeta) = \frac{(n-1)!\alpha!}{(n+|\alpha|-1)!},$$

$$(2.10) \qquad \int_B |z^\alpha|^2 \, d\nu(z) = \frac{n!\alpha!}{(n+|\alpha|)!}.$$

Proof. The proof is as given in [Ru3], but is included for completeness. Set

$$I = \int_{\mathbb{C}^n} |z^\alpha|^2 \exp(-|z|^2) \, dV(z).$$

By Fubini's Thoerem

$$I = \prod_{j=1}^n \int_{\mathbb{C}} |\lambda|^{2\alpha_j} e^{-|\lambda|^2} \, dA(\lambda).$$

Using polar coordinates,

$$\int_{\mathbb{C}} |\lambda|^{2\alpha_j} e^{-|\lambda|^2} \, dA(\lambda) = 2\pi \int_0^\infty r^{2\alpha_j+1} e^{-r^2} \, dr$$

$$= \pi \int_0^\infty t^{\alpha_j} e^{-t} \, dt$$

$$= \pi \Gamma(\alpha_j + 1) = \pi \alpha_j!.$$

Therefore $I = \pi^n \alpha!$. If we apply the integration in polar coordinates formula (1.6), we have

$$\pi^n \alpha! = 2n c_n \int_0^\infty r^{2|\alpha|+2n-1} e^{-r^2} \, dr \int_S |\zeta^\alpha|^2 \, d\sigma(\zeta)$$

$$= n c_n \Gamma(|\alpha| + n) \int_S |\zeta^\alpha|^2 \, d\sigma(\zeta).$$

Taking $\alpha = 0$ gives $c_n = \dfrac{\pi^n}{n!}$, and (2.9) now follows. To obtain (2.10), one again uses the integration in polar coordinates formula (1.6). $\square$

Returning to the example, by the above

$$|b_\alpha|^2 = \frac{(n+|\alpha|)!}{\pi^n \alpha!},$$

and thus by (2.7),

$$K(z,w) = \frac{1}{\pi^n} \sum_{k=0}^\infty \sum_{|\alpha|=k} \frac{(n+k)!}{\alpha!} z^\alpha \overline{w}^\alpha.$$

By the multinomial expansion (1.12)

$$\langle z, w \rangle^k = \sum_{|\alpha|=k} \frac{k!}{\alpha!} z^\alpha \overline{w}^\alpha.$$

Therefore

$$K(z, w) = \frac{1}{\pi^n} \sum_{k=0}^\infty \frac{(n+k)!}{k!} \langle z, w \rangle^k,$$

which by the binomial expansion (1.11) gives

$$(2.11) \qquad K(z, w) = \frac{n!}{\pi^n \left(1 - \langle z, w \rangle\right)^{n+1}}.$$

In particular, for the unit disc U,

$$K(z, w) = \frac{1}{\pi \left(1 - z\overline{w}\right)^2}.$$

(2) For the polydisc U^n, it is easily shown that

$$(2.12) \qquad K(z, w) = \frac{1}{\pi^n} \prod_{j=1}^n \frac{1}{(1 - z_j \overline{w}_j)^2}.$$

In fact, it is a relatively simple proof to shown that for a product of domains, the Bergman kernel is just the product of the kernels of the respective domains.

(3) For any integer $\alpha > 0$, $\alpha \neq 1$, let

$$D_\alpha = \{ (z_1, z_2) : |z_1|^{2/\alpha} + |z_2|^2 < 1 \}.$$

Since $V(z) = |z_1|^{2/\alpha} + |z_2|^2$ is plurisubharmonic, the domain D_α is pseudo-convex. For this domain, the Bergman kernel is given by

$$(2.13) \qquad K(z, w) = \frac{(1 - z_2 \overline{w}_2)^{\alpha - 2} \left[(\alpha + 1)(1 - z_2 \overline{w}_2)^\alpha + (\alpha - 1) z_1 \overline{w}_1 \right]}{\pi^2 \left[(1 - z_2 \overline{w}_2)^\alpha - z_1 \overline{w}_1 \right]^3}.$$

One of the difficulties with the domain D_α is that for $\alpha \neq 1$, $Aut(D_\alpha)$ is not transitive on D_α. It is known ([Be]) that the automorphisms of D_α are given by

$$\varphi(z_1, z_2) = \left(\left[\frac{\sqrt{1 - |t_2|^2}}{1 - \overline{t}_2 z_2} \right]^\alpha z_1 e^{i\theta_1}, \frac{z_2 - t_2}{1 - \overline{t}_2 z_2} e^{i\theta_2} \right),$$

where $|t_2| < 1$, $0 \leq \theta_1, \theta_2 \leq 2\pi$. It is easily seen that these are not transitive on D_α.

2.3. Properties of the Bergman Kernel.

There are several additional properties of the Bergman kernel which will be required. The first deals with the invariance under biholomorphic mappings. Let Ω_1 and Ω_2 be two domains in $\mathbb{C}^n$, and let $\psi : \Omega_1 \to \Omega_2$ be a biholomorphic mapping of Ω_1 onto Ω_2. Define

$$(2.14) \qquad A_\psi(z) = \det J_\psi(z)$$

where as in (1.3), J_ψ denotes the Jacobian matrix of the mapping ψ. Then $A_\psi(z)$ is a holomorphic function on Ω_1. By the change of variables formula,

$$(2.15) \qquad \int_{\psi(U)} f(z)\,dV(z) = \int_U f(\psi(z))|A_\psi(z)|^2\,dV(z)$$

for every Borel subset U of Ω_1 and integrable f on Ω_2. For $f \in A^2(\Omega_2)$, define f_ψ on Ω_1 by

$$f_\psi(z) = f(\psi(z))\,A_\psi(z).$$

Then f_ψ is holomorphic on Ω_1, and by (2.15) $f \to f_\psi$ is an isometry of $A^2(\Omega_2)$ to $A^2(\Omega_1)$, which is easily shown to be onto. By using the expansion (2.7) of the Bergman kernel one obtains the following:

Proposition 2.5. *Let Ω_1, Ω_2 be domains in $\mathbb{C}^n$, and let ψ be a biholomorphic mapping of Ω_1 onto Ω_2. Then*

$$(2.16) \qquad K_{\Omega_1}(z,w) = K_{\Omega_2}(\psi(z),\psi(w))A_\psi(z)\overline{A_\psi(w)},$$

where K_{Ω_i} denotes the Bergman kernel of Ω_i. In particular, if $\psi \in Aut(\Omega)$,

$$(2.17) \qquad K(z,w) = K(\psi(z),\psi(w))A_\psi(z)\overline{A_\psi(w)}.$$

One important consequence of the above concerns the invariant measure on Ω. Define λ_Ω on Ω by

$$(2.18) \qquad d\lambda_\Omega(w) = K(w,w)\,dV(w).$$

Then by (2.15) and (2.17), if $\psi \in Aut(\Omega)$ and U is a measurable subset of Ω,

$$\int_{\psi(U)} f(z)K(z,z)\,dV(z) = \int_U f(\psi(z))K(\psi(z),\psi(z))|A_\psi(z)|^2\,dV(z)$$

$$= \int_U f(\psi(z))K(z,z)\,dV(z).$$

Thus for all $\psi \in Aut(\Omega)$,

$$(2.19) \qquad \int_{\psi(U)} f(z)\,d\lambda_\Omega(z) = \int_U f(\psi(z))\,d\lambda_\Omega(z).$$

For the unit ball B, by (2.11)

$$(2.20) \qquad d\lambda_B(z) = \frac{n!}{\pi^n}\frac{dV(z)}{(1-|z|^2)^{n+1}} = \frac{d\nu(z)}{(1-|z|^2)^{n+1}}.$$

Our second result is the following:

Proposition 2.6. *Let K be the Bergman kernel of a domain Ω in $\mathbb{C}^n$. Then*

$$(2.21) \qquad \sum_{i,j=1}^{n} \frac{\partial^2 \log K(z,z)}{\partial z_i \partial \overline{z}_j} w_i \overline{w}_j > 0, \qquad z \in \Omega,\; w \in \mathbb{C}^n,\; w \neq 0.$$

Proof. The proof of this result can be obtained from the expansion (2.7) by showing that

$$\sum_{i,j=1}^{n} \frac{\partial^2 \log K(z,z)}{\partial z_i \partial \overline{z}_j} w_i \overline{w}_j \geq \frac{1}{K^2} \sum_{\beta > \alpha} \left| \sum_{j=1}^{n} \left(\psi_\alpha \frac{\partial \psi_\beta}{\partial z_j} - \psi_\beta \frac{\partial \psi_\alpha}{\partial z_j} \right) w_j \right|^2,$$

from which the result then follows. The details are omitted.

As a consequence of (1.5) and the above, $\log K(z,z)$ is plurisubharmonic on Ω. Furthermore, since

$$K \sum_{i,j=1}^{n} \frac{\partial^2 \log K(z,z)}{\partial z_i \partial \overline{z}_j} w_i \overline{w}_j = \sum_{i,j=1}^{n} \frac{\partial^2 K}{\partial z_i \partial \overline{z}_j} w_i \overline{w}_j - \frac{1}{K} \left| \sum_{j=1}^{n} \frac{\partial K}{\partial z_j} w_j \right|^2,$$

we also have

$$(2.22) \qquad \sum_{i,j=1}^{n} \frac{\partial^2 K}{\partial z_i \partial \overline{z}_j} w_i \overline{w}_j \geq \frac{1}{K} \left| \sum_{j=1}^{n} \frac{\partial K}{\partial z_j} w_j \right|^2.$$

2.4. The Bergman Metric.

For $z \in \Omega$, define

$$(2.23) \qquad g_{i,j}(z) = \frac{\partial^2 \log K(z,z)}{\partial z_i \partial \overline{z}_j}.$$

By (2.21), the hermitian form $\sum_{i,j} g_{i,j}(z) w_i \overline{w}_j$ is positive definite on $\mathbb{C}^n$ and thus defines a metric in Ω as follows: for $z \in \Omega$, $\zeta \in \mathbb{C}^n$ set

$$(2.24) \qquad \beta_\Omega(z,\zeta) = \left(\sum_{i,j} g_{i,j}(z) \zeta_i \overline{\zeta}_j \right)^{\frac{1}{2}}.$$

Then $\beta_\Omega(z,\zeta)$ is the length of the vector ζ in the Bergman metric on Ω. If $\gamma : [0,1] \to \Omega$ is a C^1 curve, the **Bergman length** of γ, denoted $|\gamma|_B$ is defined by

$$(2.25) \qquad |\gamma|_B = \int_0^1 \beta_\Omega(\gamma(t), \gamma'(t))\, dt.$$

Finally, for $z, w \in \Omega$, define

$$(2.26) \qquad \delta_\Omega(z,w) = \inf\{|\gamma|_B : \gamma(0) = z, \gamma(1) = w\},$$

where the infimum is taken over all C^1 curves from z to w. δ_Ω is called the **Bergman metric** on Ω. As a consequence of Proposition 2.5 we obtain the following:

Proposition 2.7. *Let Ω_1, Ω_2 be domains in $\mathbb{C}^n$, and let ψ be a biholomorphic mapping of Ω_1 onto Ω_2. Then for all z, $w \in \Omega_1$,*

$$\text{(2.27)} \qquad \delta_{\Omega_1}(z,w) = \delta_{\Omega_2}(\psi(z), \psi(w)).$$

Proof. By (2.16),

$$\log A_\psi(z) + \log K_{\Omega_2}(\psi(z), \psi(z)) + \log \overline{A_\psi(z)} = \log K_{\Omega_1}(z, z).$$

Since $A_\psi(z)$ is holomorphic and nonzero in Ω_1,

$$g_{i,j}^{\Omega_1}(z) = \sum_{k,m=1}^{n} g_{k,m}^{\Omega_2}(\psi(z)) \frac{\partial \psi_k(z)}{\partial z_i} \overline{\frac{\partial \psi_m(z)}{\partial z_j}},$$

and hence

$$\sum_{i,j} g_{i,j}^{\Omega_1}(z) w_i \overline{w}_j = \sum_{k,m} g_{k,m}^{\Omega_2}(\psi(z)) \left(J_\psi(z) w \right)_k \overline{\left(J_\psi(z) w \right)_m},$$

where $(J_\psi(z)w)_k$ denotes the k'th component of the vector $J_\psi(z)w$.

Suppose that γ is a C^1 curve in Ω_1. Set $\eta(t) = \psi(\gamma(t))$ and $\eta_k(t) = \psi_k(\gamma(t))$. Then

$$\eta_k'(t) = \sum_{j=1}^{n} \frac{\partial \psi_k(\gamma(t))}{\partial z_j} \gamma_j'(t) = (J_\psi(\gamma(t)\gamma'(t)))_k.$$

Therefore, by the above

$$\beta_{\Omega_2}(\eta(t), \eta'(t)) = \beta_{\Omega_1}(\gamma(t), \gamma'(t)),$$

from which it now follows by (2.25) that $|\gamma|_B = |\psi \circ \gamma|_B$. $\square$

We conclude this section by computing the Bergman metric for B_n. From formula (2.11), by a straightforward computation, for $n > 1$,

$$\text{(2.28)} \qquad g_{i,j}(z) = \frac{(n+1)}{(1-|z|^2)^2} \left[(1-|z|^2)\delta_{i,j} + \overline{z}_i z_j \right],$$

and thus

$$\text{(2.29)} \qquad \beta^2(z, \zeta) = \frac{(n+1)}{(1-|z|^2)^2} \left[(1-|z|^2)|\zeta|^2 + |\langle z, \zeta \rangle|^2 \right].$$

When $n = 1$ this becomes

$$g(z) = g_{1,1}(z) = \frac{2}{(1-|z|^2)^2} \qquad \text{and} \qquad \beta^2(z, \zeta) = \frac{2|\zeta|^2}{(1-|z|^2)^2}.$$

To derive the formula for the Bergman metric on B_n, one first shows that for the point re_1, $0 < r < 1$,

$$\delta_B(0, re_1) = \frac{\sqrt{n+1}}{2} \log \frac{1+r}{1-r}.$$

The term on the right is obtained by taking $\gamma(t) = tre_1$, $0 \leq t \leq 1$, and it can be shown that the infimum in (2.25) is attained for this curve. In fact, if we write

$$\zeta = P_z\zeta + Q_z\zeta$$

where $P_z\zeta$ is the orthogonal projection of ζ onto the plane $[z]$ spanned by 0 and z, then

$$(1 - |z|^2)|\zeta|^2 + |\langle z, \zeta \rangle|^2 = |P_z\zeta|^2 + (1 - |z|^2)|Q_z\zeta|^2$$
$$\geq |P_z\zeta|^2,$$

with equality if and only if $\zeta \in [z]$. Thus if γ is any curve in B,

$$\beta(\gamma(t), \gamma'(t)) \geq \sqrt{n+1} \frac{|P_{\gamma(t)}\gamma'(t)|}{1 - |\gamma(t)|^2}.$$

If we set $u(t) = |\gamma(t)|$, then $|P_{\gamma(t)}\gamma'(t)| \geq |u'(t)|$, and thus

$$\int_0^1 \beta(\gamma(t), \gamma'(t)) \geq \sqrt{n+1} \int_0^1 \frac{|u'(t)|}{1 - u^2(t)} \, dt$$
$$\geq \sqrt{n+1} \left| \int_0^1 \frac{u'(t)}{1 - u^2(t)} \, dt \right|,$$

from which the result now follows.

If $z \in B$ is arbitary, we can choose $U \in \mathcal{U}$ such that $z = rUe_1$, $r = |z|$. Thus since δ_B is invariant under $\mathcal{U}$, the above formula gives $\delta_B(0, z)$. Finally, for $z, w \in B$, let $\varphi_w \in Aut(B)$ be as defined by (1.13). Then by Proposition 2.7 ,

$$(2.30) \qquad \delta_B(w, z) = \delta_B(0, \varphi_w(z)) = \frac{\sqrt{n+1}}{2} \log \frac{1 + |\varphi_w(z)|}{1 - |\varphi_w(z)|}$$
$$= \sqrt{n+1} \tanh^{-1} |\varphi_w(z)|.$$

3.
The Laplace-Beltrami Operator

For a bounded domain Ω in $\mathbb{C}^n$, the Bergman metric gives Ω the structure of a Hermitian manifold. In this chapter we introduce both the Laplace-Beltrami operator and the gradient with respect to this Hermitian structure.

3.1. The Invariant Laplacian.

Let Ω be a bounded domain in $\mathbb{C}^n$ with Bergman kernel K, and let $(g_{i,j}(z))$ denote the $n \times n$ Hermitian matrix given by (2.23). Since the matrix $(g_{i,j})$ is positive definite, it is invertible. Let $(g^{i,j}(z))$ denote the inverse matrix, and set $g(z) = \det (g_{i,j}(z))$. The **Laplace-Beltrami** operator associated with K is the differential operator $\widetilde{\Delta}_\Omega$ defined by

$$(3.1) \qquad \widetilde{\Delta}_\Omega = \frac{2}{g} \sum_{i,j} \left\{ \frac{\partial}{\partial \overline{z}_i} \left(g\, g^{i,j} \frac{\partial}{\partial z_j} \right) + \frac{\partial}{\partial z_j} \left(g\, g^{i,j} \frac{\partial}{\partial \overline{z}_i} \right) \right\}.$$

The following result, the proof of which can be found in [He2], is valid for the Laplace-Beltrami operator on any Riemannian manifold.

Proposition 3.1.

   (1) If $f \in C^2(\Omega)$, then for all $\psi \in Aut(\Omega)$,

$$(3.2) \qquad \widetilde{\Delta}_\Omega(f \circ \psi) = (\widetilde{\Delta}_\Omega f) \circ \psi.$$

   (2) If $f, g \in C_c^2(\Omega)$, then

$$(3.3) \qquad \int_\Omega f\, \widetilde{\Delta}_\Omega g\, d\lambda_\Omega = \int_\Omega g\, \widetilde{\Delta}_\Omega f\, d\lambda_\Omega.$$

Property (3.2) is the invariance property of the operator $\widetilde{\Delta}_\Omega$, and for this reason $\widetilde{\Delta}_\Omega$ is often referred to as the **invariant** Laplacian. Identity (3.3) is Green's formula for the operator $\widetilde{\Delta}_\Omega$.

In the next two sections we examine the Laplace-Beltrami operator both on the polydisc U^n and the unit ball B_n.

3.2. The Invariant Laplacian for U^n.

For the polydisc U^n, as a consequence of (2.12)

$$g_{i,j}(z) = \frac{2\,\delta_{i,j}}{(1 - |z_j|^2)^2},$$

and thus

$$g^{i,j}(z) = \frac{1}{2}(1 - |z_j|^2)^2 \delta_{i,j},$$

and

$$g(z) = 2^n \prod_{j=1}^{n} \frac{1}{(1 - |z_j|^2)^2}.$$

Remark: It should be noted that in the above

$$g(z) = c\,K(z,z),$$

where c is a positive constant. This identity is in fact valid for any domain ([He1]).

Since $g^{i,j} = 0$ for $i \neq j$,

$$\tilde{\Delta}_{U^n} = \frac{2}{g} \sum_{j=1}^{n} \left\{ \frac{\partial}{\partial \bar{z}_j} \left(g\,g^{j,j} \frac{\partial}{\partial z_j} \right) + \frac{\partial}{\partial z_j} \left(g\,g^{j,j} \frac{\partial}{\partial \bar{z}_j} \right) \right\}.$$

In this example, $g(z)g^{j,j}(z) = 2^{n-1} \prod_{i \neq j}(1 - |z_i|^2)^{-2}$, and thus

$$\frac{\partial}{\partial \bar{z}_j} g\,g^{j,j} = \frac{\partial}{\partial z_j} g\,g^{j,j} = 0.$$

Therefore

(3.4)
$$\tilde{\Delta}_{U^n} = 2 \sum_{j=1}^{n}(1 - |z_j|^2)^2 \frac{\partial^2}{\partial z_j \partial \bar{z}_j} = \sum_{j=1}^{n} \tilde{\Delta}_j,$$

where

$$\tilde{\Delta}_j = 2(1 - |z_j|^2)^2 \frac{\partial^2}{\partial z_j \partial \bar{z}_j}.$$

Recall from the introduction that a function f on U^n is **n-harmonic** (or strongly harmonic) if it is harmonic in each variable separately. Thus a C^2 function is n-harmonic on U^n if and only if $\tilde{\Delta}_j f = 0$ for all $j = 1, ..., n$. We now give an example of a function which satisfies $\tilde{\Delta}_{U^n} f = 0$, but which is not n-harmonic on U^n. For $z \in U$, $t \in T$, let

(3.5)
$$P(z,t) = \frac{(1 - |z|^2)}{|1 - z\,\bar{t}|^2}$$

denote the Poisson kernel on U. As we will see in Proposition 5.4, for fixed $t \in T$, $\alpha > 0$,

$$\widetilde{\Delta}_U P^\alpha(z,t) = 2\alpha(\alpha - 1)P^\alpha(z,t).$$

For $i = 1, 2$, let $\alpha_i = \lambda_1 + \frac{1}{2}$. Then

$$\widetilde{\Delta}_i P^{\frac{1}{2}+\lambda_i}(z_i, t_i) = 2(\lambda_i^2 - \tfrac{1}{4})P^{\frac{1}{2}+\lambda_i}(z_i, t_i).$$

Thus if $F(z_1, z_2) = P^{\frac{1}{2}+\lambda_1}(z_1, t_1)P^{\frac{1}{2}+\lambda_2}(z_2, t_2)$,

$$\widetilde{\Delta}_{U^2} F = 2(\lambda_1^2 + \lambda_2^2 - \tfrac{1}{2})F.$$

Hence for any λ_1, λ_2 satisfying $\lambda_1^2 + \lambda_2^2 = \frac{1}{2}$, $\widetilde{\Delta}_{U^2} F = 0$. If λ_1 and λ_2 are not equal to $\frac{1}{2}$, the function F is not 2-harmonic. Thus for an appropriate choice of λ_1, λ_2, F is weakly harmonic but not strongly harmonic.

3.3. The Invariant Laplacian for B.

For the unit ball B_n it was shown in (2.28) that

$$g_{i,j}(z) = \frac{(n+1)}{(1-|z|^2)^2}\left[(1-|z|^2)\delta_{i,j} + \bar{z}_i z_j\right].$$

The inverse matrix $(g^{i,j}(z))$ is given by

(3.6)
$$g^{i,j}(z) = \frac{(1-|z|^2)}{n+1}\left[\delta_{i,j} - \bar{z}_i z_j\right],$$

and

$$g(z) = \frac{(n+1)^n}{(1-|z|^2)^{n+1}}.$$

We now show that

(3.7)
$$\widetilde{\Delta}_B = \frac{4(1-|z|^2)}{n+1}\sum_{i,j=1}^{n}\left[\delta_{i,j} - \bar{z}_i z_j\right]\frac{\partial^2}{\partial z_j \partial \bar{z}_i}.$$

By (3.1),

$$\widetilde{\Delta}_B = 4\sum_{i,j=1}^{n} g^{i,j}\frac{\partial^2}{\partial z_j \partial \bar{z}_i} + \frac{2}{g}\sum_{i,j=1}^{n}\frac{\partial}{\partial \bar{z}_i}(g\,g^{i,j})\frac{\partial}{\partial z_j} + \frac{2}{g}\sum_{i,j=1}^{n}\frac{\partial}{\partial z_j}(g\,g^{i,j})\frac{\partial}{\partial \bar{z}_i}.$$

Since $g(z)g^{i,j}(z) = (n+1)^{n-1}[\delta_{i,j} - \bar{z}_i z_j](1-|z|^2)^{-n}$,

$$\frac{\partial}{\partial \bar{z}_i}(g(z)g^{i,j}(z)) = (n+1)^{n-1}\frac{[nz_i(\delta_{i,j} - \bar{z}_i z_j) - z_j(1-|z|^2)]}{(1-|z|^2)^{n+1}}.$$

Thus for fixed j,

$$\sum_{i=1}^{n} \frac{\partial}{\partial \bar{z}_i} g\, g^{i,j} = \frac{(n+1)^{n-1}}{(1-|z|^2)^{n+1}} \left[nz_j - nz_j \sum_{i=1}^{n} |z_i|^2 - nz_j(1-|z|^2) \right] = 0.$$

Therefore

$$\sum_{j=1}^{n} \left(\sum_{i=1}^{n} \frac{\partial}{\partial \bar{z}_i} g\, g^{i,j} \right) \frac{\partial}{\partial z_j} = 0.$$

A similar computation shows that the other term vanishes likewise, thus leaving (3.7).

Remark: From formula (3.7) we obtain that for a C^2 function f,

$$(\tilde{\Delta}_B f)(0) = \frac{4}{n+1} \sum_{j=1}^{n} \frac{\partial^2 f}{\partial z_j \partial \bar{z}_j}(0) = \frac{1}{n+1}(\Delta f)(0)$$

where

$$\Delta = 4 \sum_{j=1}^{n} \frac{\partial^2}{\partial z_j \partial \bar{z}_j}$$

is the ordinary Laplacian on $\mathbb{C}^n$ ($\mathbb{R}^{2n}$). Thus by the invariance property (3.2), if f is C^2 on B,

(3.8) $$(\tilde{\Delta}_B f)(a) = \tfrac{1}{n+1} \Delta(f \circ \varphi_a)(0).$$

For future reference, we also include the radial form of the invariant Laplacian on B. A function f on B is **radial** if $f(z) = f(|z|)$ for all $z \in B$. Suppose f is a C^2 radial function on B. Then with $r = |z|$,

$$\frac{\partial^2 f}{\partial z_j \partial \bar{z}_i} = f''(r) \frac{\partial r}{\partial \bar{z}_i} \frac{\partial r}{\partial z_j} + f'(r) \frac{\partial^2 r}{\partial \bar{z}_i \partial z_j}.$$

Since $r^2 = \sum_{i=1}^{n} z_i \bar{z}_i$,

$$\frac{\partial r}{\partial \bar{z}_i} = \frac{z_i}{2r}, \quad \frac{\partial r}{\partial z_j} = \frac{\bar{z}_j}{2r}, \quad \frac{\partial^2 r}{\partial z_j \partial \bar{z}_i} = \frac{2r^2 \delta_{i,j} - z_i \bar{z}_j}{4r^3}.$$

Substituting into (3.7) and summing the terms gives

(3.9) $$\tilde{\Delta}_B f(z) = \frac{(1-r^2)}{n+1} \left[(1-r^2)f''(r) + \frac{(2n - r^2 - 1)}{r} f'(r) \right].$$

There is one additional characterization of $\tilde{\Delta}_B$ which will prove useful.

Proposition 3.2. If $f \in C^2(B)$, then for all $a \in B$,

$$(3.10) \qquad (\tilde{\Delta}_B f)(a) = \lim_{r \to 0^+} \frac{4n}{(n+1)r^2} \int_S [f(\varphi_a(rt)) - f(a)] \, d\sigma(t).$$

Proof. Let $h = f \circ \varphi_a$. Taking the Taylor expansion of h about 0 gives

$$h(z) = h(0) + \sum_{j=1}^{n} \left(z_j \frac{\partial h}{\partial z_j}(0) + \bar{z}_j \frac{\partial h}{\partial \bar{z}_j}(0) \right) + \frac{1}{2} \sum_{j=1}^{n} \left(z_j^2 \frac{\partial^2 h}{\partial z_j^2}(0) + \bar{z}_j^2 \frac{\partial^2 h}{\partial \bar{z}_j^2}(0) \right)$$

$$+ \sum_{i,j=1}^{n} z_i \bar{z}_j \frac{\partial^2 h}{\partial z_i \partial \bar{z}_j}(0) + O(|z|^3).$$

Integrating $h(rt)$ over S and using formula (1.7) gives

$$\int_S h(rt) \, d\sigma(t) = h(0) + \sum_{j=1}^{n} \frac{\partial^2 h}{\partial z_j \partial \bar{z}_j}(0) \, r^2 \int_S |t_j|^2 \, d\sigma(t) + O(r^3).$$

Thus by (2.9),

$$\frac{4n}{r^2} \int_S [h(rt) - h(0)] \, d\sigma(t) = \Delta h(0) + O(r).$$

The result now follows by (3.8). $\square$

3.4. The Invariant Gradient.

We close this chapter with a discussion of the **gradient** with respect to the Bergman metric on Ω. If u and v are C^2 functions on Ω, then by (3.1)

$$(3.11) \qquad \tilde{\Delta}(u\,v) = u\,\tilde{\Delta}v + 2(\tilde{\nabla}u)v + v\,\tilde{\Delta}u,$$

where now for a C^1 function u, $\tilde{\nabla}u$ is the vector field on Ω defined by

$$(3.12) \qquad \tilde{\nabla}u = 2 \sum_{i,j} g^{i,j} \left\{ \frac{\partial u}{\partial \bar{z}_i} \frac{\partial}{\partial z_j} + \frac{\partial u}{\partial z_j} \frac{\partial}{\partial \bar{z}_i} \right\}.$$

From the above it is clear that $(\tilde{\nabla}u)v = (\tilde{\nabla}v)u$. If u and v are real valued, then

$$\frac{\partial u}{\partial \bar{z}_i} = \overline{\frac{\partial u}{\partial z_i}} \qquad \text{and} \qquad \frac{\partial v}{\partial \bar{z}_j} = \overline{\frac{\partial v}{\partial z_j}},$$

and thus

$$\begin{aligned}
(\tilde{\nabla}u)v &= 2\sum_{i,j} g^{i,j}\frac{\partial u}{\partial \bar{z}_i}\frac{\partial v}{\partial z_j} + 2\sum_{i,j} g^{j,i}\frac{\partial u}{\partial z_i}\frac{\partial v}{\partial \bar{z}_j} \\
&= 2\sum_{i,j}\left\{g^{i,j}\frac{\overline{\partial u}}{\partial z_i}\frac{\partial v}{\partial z_j} + \overline{g^{i,j}\frac{\partial u}{\partial z_i}}\frac{\partial v}{\partial z_j}\right\} \\
&= 4\,\mathrm{Re}\left(\sum_{i,j} g^{i,j}\frac{\overline{\partial u}}{\partial z_i}\frac{\partial v}{\partial z_j}\right).
\end{aligned}$$

In the above we have used the fact that $g^{j,i} = \overline{g^{i,j}}$. Since the matrix $(g^{i,j})$ is positive definite, $(\tilde{\nabla}u)u \geq 0$.

If we take $v = u$ in (3.11), then

$$(3.13) \qquad \tilde{\Delta}u^2 = 2u\,\tilde{\Delta}u + 2\,|\tilde{\nabla}u|^2,$$

where for real valued u,

$$(3.14) \qquad |\tilde{\nabla}u|^2 = (\tilde{\nabla}u)u = 4\sum_{i,j} g^{i,j}\frac{\overline{\partial u}}{\partial z_i}\frac{\partial u}{\partial z_j}.$$

Since $\tilde{\Delta}$ is invariant under $Aut(\Omega)$, by (3.2) and the above,

$$(3.15) \qquad |\tilde{\nabla}(u\circ\psi)|^2 = |(\tilde{\nabla}u)\circ\psi|^2$$

for all $\psi \in Aut(\Omega)$.

In the case of the unit ball, for real valued u

$$(3.16)\qquad \begin{aligned}
|\tilde{\nabla}u(z)|^2 &= \frac{4}{n+1}(1-|z|^2)\sum_{i,j}^{n}(\delta_{i,j}-\bar{z}_iz_j)\frac{\partial u}{\partial \bar{z}_i}\frac{\partial u}{\partial z_j} \\
&= \frac{4}{n+1}(1-|z|^2)\left[\sum_{j=1}^{n}\left|\frac{\partial u}{\partial z_j}\right|^2 - \left|\sum_{j=1}^{n}z_j\frac{\partial u}{\partial z_j}\right|^2\right].
\end{aligned}$$

In particular,

$$|\tilde{\nabla}u(0)|^2 = \frac{4}{n+1}\sum_{j=1}^{n}\frac{\partial u}{\partial \bar{z}_j}(0)\frac{\partial u}{\partial z_j}(0) = \frac{1}{n+1}|\nabla u(0)|^2,$$

where

$$\nabla = \left(\frac{\partial}{\partial x_1},\ \ldots,\ \frac{\partial}{\partial x_n},\ \frac{\partial}{\partial y_i},\ \ldots,\ \frac{\partial}{\partial y_n}\right).$$

is the usual gradient on $\mathbb{R}^{2n}$. Therefore by (3.15),

$$|\widetilde{\nabla}u(z)|^2 = \frac{1}{n+1}|\nabla(u \circ \varphi_z)(0)|^2.$$

There is a natural way to define an inner product on the gradient vector fields which will justify the use of the notation $|\widetilde{\nabla}u|^2$ introduced in (3.14). The vector fields

$$\left\{ \frac{\partial}{\partial z_1}, \frac{\partial}{\partial \bar{z}_1}, \ldots, \frac{\partial}{\partial z_n}, \frac{\partial}{\partial \bar{z}_n} \right\}$$

form a basis (over the reals) for the complex tangent space, and we define a hermitian inner product (with respect to the Bergman metric) on the complexification of the tangent space at a point $z \in \Omega$ by

$$\left\langle \frac{\partial}{\partial z_i}, \frac{\partial}{\partial z_j} \right\rangle = \left\langle \frac{\partial}{\partial \bar{z}_j}, \frac{\partial}{\partial \bar{z}_i} \right\rangle = \frac{1}{2}g_{i,j}(z),$$

and

$$\left\langle \frac{\partial}{\partial z_i}, \frac{\partial}{\partial \bar{z}_j} \right\rangle = \left\langle \frac{\partial}{\partial \bar{z}_i}, \frac{\partial}{\partial z_j} \right\rangle = 0.$$

If u is C^1, then

$$\left\langle \widetilde{\nabla}u, \frac{\partial}{\partial z_\alpha} \right\rangle = 2\sum_{i,j} g^{i,j} \frac{\partial u}{\partial \bar{z}_i} \left\langle \frac{\partial}{\partial z_j}, \frac{\partial}{\partial z_\alpha} \right\rangle$$

$$= \sum_{i,j} \frac{\partial u}{\partial \bar{z}_i} g^{i,j} g_{j,\alpha}$$

$$= \frac{\partial u}{\partial \bar{z}_\alpha}.$$

In the above we have used the fact that

$$\sum_{j=1}^{n} g^{i,j} g_{j,\alpha} = \delta_{i,\alpha}.$$

Similarly,

$$\left\langle \widetilde{\nabla}u, \frac{\partial}{\partial \bar{z}_\alpha} \right\rangle = \frac{\partial u}{\partial z_\alpha}.$$

Thus by linearity, if $u, v \in C^1(\Omega)$,

$$\langle \widetilde{\nabla}u, \widetilde{\nabla}v \rangle = 2\sum_{i,j} \overline{g^{i,j}} \left\{ \overline{\frac{\partial v}{\partial \bar{z}_i}} \left\langle \operatorname{grad} u, \frac{\partial}{\partial z_j} \right\rangle + \overline{\frac{\partial v}{\partial z_j}} \left\langle \operatorname{grad} u, \frac{\partial}{\partial \bar{z}_i} \right\rangle \right\}$$

$$= 2\sum_{i,j} g^{i,j} \left\{ \frac{\partial u}{\partial \bar{z}_i} \frac{\partial \bar{v}}{\partial z_j} + \frac{\partial u}{\partial z_j} \frac{\partial \bar{v}}{\partial \bar{z}_i} \right\}$$

$$= (\widetilde{\nabla}u)\bar{v}.$$

In the above we have used that for complex valued v, $\dfrac{\overline{\partial v}}{\partial \bar{z}_i} = \dfrac{\partial \bar{v}}{\partial z_i}$ and $\dfrac{\overline{\partial v}}{\partial z_i} = \dfrac{\partial \bar{v}}{\partial \bar{z}_i}$. In particular, for a C^1 function f,

$$(3.17) \quad |\widetilde{\nabla} f|^2 = \langle \widetilde{\nabla} f, \widetilde{\nabla} f \rangle = (\widetilde{\nabla} f)\bar{f} = 2 \sum_{i,j} g^{i,j} \left\{ \frac{\partial f}{\partial \bar{z}_i} \frac{\overline{\partial f}}{\partial \bar{z}_j} + \frac{\overline{\partial f}}{\partial z_i} \frac{\partial f}{\partial z_j} \right\},$$

which is nonnegative since the matrix $(g^{i,j})$ is positive definite. In particular, if f is **holomorphic** on Ω, then

$$(3.18) \qquad |\widetilde{\nabla} f|^2 = 2 \sum_{i,j} g^{i,j} \frac{\overline{\partial f}}{\partial z_i} \frac{\partial f}{\partial z_j},$$

and thus for the unit ball,

$$(3.19) \qquad |\widetilde{\nabla} f(z)|^2 = \frac{2}{n+1}(1 - |z|^2) \left[\sum_{j=1}^{n} \left| \frac{\partial f}{\partial z_j} \right|^2 - \left| \sum_{j=1}^{n} z_j \frac{\partial f}{\partial z_j} \right|^2 \right],$$

which (except for a factor of 2) corresponds to formula (3.16) for $|\widetilde{\nabla} u|^2$ for a real valued function u. When $n = 1$, the above becomes

$$|\widetilde{\nabla} f(z)|^2 = (1 - |z|^2)^2 |f'(z)|^2.$$

Furthermore, if f is holomorphic, then by (3.11)

$$(3.20) \qquad \widetilde{\Delta} |f|^2 = 2(\widetilde{\nabla} f)\bar{f} = 2|\widetilde{\nabla} f|^2.$$

Also, since $|\widetilde{\nabla}(f \circ \psi)|^2 = |(\widetilde{\nabla} f) \circ \psi|^2$ for all $\psi \in Aut(\Omega)$, for holomorphic functions f on B we have

$$|(\widetilde{\nabla} f)(z)|^2 = \frac{2}{n+1} |\partial(f \circ \varphi_z)(0)|^2,$$

where $\partial = \left(\frac{\partial}{\partial z_1}, ..., \frac{\partial}{\partial z_n} \right)$.

4.
Invariant Harmonic and Subharmonic Functions

As was indicated in the introduction, a C^2 function f on a bounded domain Ω is **weakly harmonic** on Ω if $\widetilde{\Delta}_\Omega f = 0$. Similary, f is said to be **weakly subharmonic** on Ω if $\widetilde{\Delta}_\Omega f \geq 0$. In the case of the unit ball, since the concept of weakly harmonic and strongly harmonic coincide, we will simply refer to such functions as **invariant** harmonic or $\mathcal{M}$-**harmonic** functions. Throughout this chapter we will simply use the notation $\widetilde{\Delta}$ and λ to denote the invariant Laplacian and invariant measure on B.

4.1. $\mathcal{M}$-Subharmonic Functions.

Rather than defining $\mathcal{M}$-harmonic and $\mathcal{M}$-subharmonic functions in terms of the invariant Laplacian, we will make the following definition, which as we will subsequently see is equivalent for C^2 functions.

Definition. *An upper semicontinuous function* $f : B \rightarrow [-\infty, \infty)$, *with* $f \not\equiv -\infty$, *is* $\mathcal{M}$-**subharmonic** *on* B *if*

$$(4.1) \qquad\qquad f(a) \leq \int_S f(\varphi_a(rt))\, d\sigma(t)$$

for all $a \in B$ *and all* r *sufficiently small. A continuous function* f *for which equality holds in (4.1) is said to be* $\mathcal{M}$-**harmonic** *on* B. *A function* f *is* $\mathcal{M}$-**superharmonic** *if* $-f$ *is* $\mathcal{M}$-*subharmonic.*

Inequality (4.1) is the invariant mean value inequality. The local nature of the definition allows one to define $\mathcal{M}$-subharmonic and $\mathcal{M}$-harmonic functions on open subsets of B in the obvious way. As we will see in Proposition 5.11, if f is $\mathcal{M}$-subharmonic on B, then for every $a \in B$, $\int_S f(\varphi_a(rt))d\sigma(t)$ is a nondecreasing function of r, and thus (4.1) holds for all r, $0 < r < 1$. We will use this fact in the following proposition and also in Proposition 4.7.

As a consequence of (3.10), a C^2 $\mathcal{M}$-subharmonic function f satisfies $(\widetilde{\Delta}f)(a) \geq 0$ for all $a \in B$, and if f is $\mathcal{M}$-harmonic, then $(\widetilde{\Delta}f)(a) = 0$ for all $a \in B$. We now prove the converse.

Proposition 4.1. *If $f \in C^2(B)$ satisfies $\widetilde{\Delta} f \geq 0$, then f is $\mathcal{M}$-subharmonic on B.*

Proof. Suppose first that f satisfies $\widetilde{\Delta} f > 0$ on B. Then as a consequence of Proposition 3.2, for each $a \in B$, there exists $r_a > 0$ such that

$$\int_S [f(\varphi_a(rt)) - f(a)] \, d\sigma(t) \geq 0$$

for all r, $0 < r < r_a$. Thus f is $\mathcal{M}$-subharmonic on B. For arbitrary f let

$$f_\epsilon(z) = f(z) + \epsilon |z|^2, \qquad \epsilon > 0.$$

Then by (3.9),

$$\widetilde{\Delta} f_\epsilon(z) = (\widetilde{\Delta} f)(z) + \frac{4\epsilon}{n+1}(1 - |z|^2)(n - |z|^2),$$

and thus $\widetilde{\Delta} f_\epsilon(z) > 0$ for all $z \in B$. By the above, f_ϵ is $\mathcal{M}$-subharmonic on B. Therefore,

$$\int_S f_\epsilon(\varphi_a(rt)) \, d\sigma(t) \geq f_\epsilon(a) \geq f(a).$$

Since the above holds for all r, $0 < r < 1$, letting $\epsilon \to 0^+$ shows that f satisfies (4.1) and thus is $\mathcal{M}$-subharmonic on B. □

If f is a C^2 $\mathcal{M}$-subharmonic function, then by Propositions 3.1 and 4.1 it is clear that $f \circ \psi$ is $\mathcal{M}$-subharmonic for all $\psi \in \mathcal{M}$. This result holds in general.

Proposition 4.2. *If f is $\mathcal{M}$-subharmonic on B, then $f \circ \psi$ is $\mathcal{M}$-subharmonic for all $\psi \in \mathcal{M}$.*

Proof. Let $a \in B$ and let $b = \psi(a)$. Then $(\varphi_b \circ \psi \circ \varphi_a)(0) = 0$, and thus by Cartan's theorem $\varphi_b \circ \psi \circ \varphi_a = U \in \mathcal{U}$. Therefore $\psi(\varphi_a(z)) = \varphi_b(Uz)$. Consequently, by the $\mathcal{U}$ invariance of σ,

$$\int_S (f \circ \psi)(\varphi_a(rt)) \, d\sigma(t) = \int_S f(\varphi_b(Urt)) \, d\sigma(t)$$

$$= \int_S f(\varphi_b(rt)) \, d\sigma(t) \geq f(b) = (f \circ \psi)(a). □$$

We now derive an invariant volume version of (4.1) and also (3.10). Since

$$\lambda(B(0,r)) = \int_{B(0,r)} \frac{d\nu(z)}{(1 - |z|^2)^{n+1}} = 2n \int_0^r \frac{\rho^{2n-1} \, d\rho}{(1 - \rho^2)^{n+1}},$$

if we multiply both sides of (4.1) by $2n\,\rho^{2n-1}(1-\rho^2)^{-n-1}$ and integrate, we obtain

$$f(a) \le \frac{1}{\lambda(B(0,r))} \int_{B(0,r)} f(\varphi_a(z))\,d\lambda(z).$$

Thus if

(4.2) $$E(a,r) = \varphi_a(B(0,r)) = \{w : |\varphi_a(w)| < r\},$$

by the change of variable formula (2.19),

(4.3) $$f(a) \le \frac{1}{\lambda(E(a,r))} \int_{E(a,r)} f(w)\,d\lambda(w)$$

for all r sufficiently small, with equality if f is $\mathcal{M}$-harmonic.

From inequality (4.3) it now follows as in the classical case that $\mathcal{M}$-subharmonic functions are locally integrable on B, i.e., $\int_K f\,d\lambda > -\infty$ for every compact set K, and satisfy the following maximum principle.

Proposition 4.3. *(Maximum Principle) Let Ω be an open connected subset of B. If f is a nonconstant $\mathcal{M}$-subharmonic function on Ω, then*

$$f(z) < \sup_{w\in\Omega} f(w), \qquad \text{for all} \quad z \in \Omega.$$

Proof. The proof of both the maximum principle and local integrability are the same as in the classical case using inequality (4.3) and a connectivity argument (e.g. [Hel]).

For completeness, we now show that

(4.4) $$\lambda(E(a,r)) = \frac{r^{2n}}{(1-r^2)^n}.$$

As above,

$$\lambda(E(a,r)) = 2n \int_0^r \frac{\rho^{2n-1}\,d\rho}{(1-\rho^2)^{n+1}}.$$

By the change of variable $\rho = \tanh t$,

$$2n \int_0^r \frac{\rho^{2n-1}\,d\rho}{(1-\rho^2)^{n+1}} = 2n \int_0^{\tanh^{-1} r} (\sinh t)^{2n-1}\,\cosh t\,dt$$

$$= \left[\sinh(\tanh^{-1} r)\right]^{2n} = \frac{r^{2n}}{(1-r^2)^n}.$$

The invariant volume version of (3.10) is obtained as follows. As in Proposition (3.2) let $h(z) = (f \circ \varphi_a)(z)$. Then

$$\int_{B(0,r)} [h(z) - h(0)]\, d\lambda(z) = 2n \int_0^r \frac{\rho^{2n-1}}{(1-\rho^2)^{n+1}} \frac{\rho^2}{4n} \Delta h(0)\, dr + O(r^{2n+3})$$

$$= \frac{c(r)}{2} \Delta h(0) + O(r^{2n+3}),$$

where $c(r) = \int_0^r \rho^{2n+1}(1-\rho^2)^{-n-1}\, d\rho$. By L'Hopitals rule,

$$\lim_{r \to 0} \frac{r^2\, \lambda(B(0,r))}{c(r)} = 2(n+1).$$

Therefore

(4.5) $(\widetilde{\Delta} f)(a) = \lim_{r \to 0} \frac{4}{r^2\, \lambda(E(a,r))} \int_{E(a,r)} [f(w) - f(a)]\, d\lambda(w).$

Remark. When $n = 1$, $E(a,r)$ is the noneuclidean disc with noneuclidean center a and radius r. When $n > 1$ ([Ru3])

(4.6) $E(a,r) = \left\{ z \in B : \frac{|P_a z - c|^2}{r^2 \rho^2} + \frac{|Q_a z|^2}{r^2 \rho} < 1 \right\},$

where

(4.7) $c = \frac{(1-r^2)\, a}{1 - r^2 |a|^2}$ and $\rho = \frac{1 - |a|^2}{1 - r^2 |a|^2}.$

The intersection of $E(a,r)$ with the plane $[a]$ is a disc with center c and radius $r\rho$, whereas it's intersection with the real $(2n - 2)$-dimensional space perpendicular to $[a]$ at c is a ball with larger radius $r\sqrt{\rho}$.

4.2. The Invariant Convolution on B.

For $0 < p < \infty$, $L^p(B, \lambda)$ denotes the space of measurable functions on B for which

$$\|f\|_p^p = \int_B |f(w)|^p\, d\lambda(w) < \infty.$$

Also, $L_{loc}^p(B)$ will denote the space of measurable functions on B which are locally p-integrable, i.e.,

$$\int_K |f(w)|^p\, d\lambda(w) < \infty$$

for every compact subset K of B. For measurable functions f, g, we define the invariant convolution of f and g by

(4.8) $(f * g)(z) = \int_B f(w) g(\varphi_z(w))\, d\lambda(w), \quad z \in B,$

provided this integral exists. By the invariance of λ we have $(f * g)(z) = (g * f)(z)$. Although the convolution as defined is not the usual definition for convolution of functions on a topological group, the following analogues of the standard convolution inequalities are still valid ([HSY]).

Lemma 4.4. *Let $p \in [1, +\infty)$ and let p' be defined by $\frac{1}{p} + \frac{1}{p'} = 1$. If $f \in L^p(B, \lambda)$, then*

$$(4.9) \qquad \|f * g\|_p \leq \|f\|_p \|g\|_1$$

for all radial functions $g \in L^1(B, \lambda)$, and

$$(4.10) \qquad \|f * g\|_\infty \leq \|f\|_p \|g\|_{p'}$$

for all radial functions $g \in L^{p'}(B, \lambda)$.

Proof. Let $g \in L^1(B, \lambda)$ and $h \in L^{p'}(B, \lambda)$, which without loss of generality may be assumed to be nonnegative. Thus by the definition of the convolution and Tonelli's theorem,

$$\int_B h(z)(f * g)(z) \, d\lambda(z) = \int_B h(z) \int_B f(w)(g \circ \varphi_z)(w) \, d\lambda(w) \, d\lambda(z)$$

$$= \int_B f(w) \int_B h(z)(g \circ \varphi_z)(w) \, d\lambda(z) \, d\lambda(w).$$

By (1.13) $|\varphi_z(w)| = |\varphi_w(z)|$. Thus since g is radial,

$$\int_B h(z)g(\varphi_z(w)) \, d\lambda(z) = \int_B h(z)g(\varphi_w(z)) \, d\lambda(z) = \int_B g(z)h(\varphi_w(z)) \, d\lambda(z).$$

The last equality follows by the $\mathcal{M}$-invariance of λ. By two successive applications of Hölder's inequality we obtain

$$\|(f * g)h\|_1 \leq \|g\|_1^{\frac{1}{p}} \int_B f(w) \left\{ \int_B g(z)(h \circ \varphi_w)^{p'}(z)d\lambda(z) \right\}^{\frac{1}{p'}} d\lambda(w)$$

$$\leq \|g\|_1^{\frac{1}{p}} \|f\|_p \left\{ \int_B \int_B g(z)(h \circ \varphi_w)^{p'}(z) \, d\lambda(z) \, d\lambda(w) \right\}^{\frac{1}{p'}}.$$

As above, since λ is invariant and g is radial,

$$\int_B \int_B g(z)(h \circ \varphi_w)^{p'}(z) \, d\lambda(z) \, d\lambda(w) = \int_B \int_B (g \circ \varphi_w)(z)h^{p'}(z) \, d\lambda(z) \, d\lambda(w)$$

$$= \int_B \int_B (g \circ \varphi_z)(w)h^{p'}(z) \, d\lambda(w) \, d\lambda(z)$$

$$= \int_B h^{p'}(z) \, d\lambda(z) \int_B g(w) \, d\lambda(w).$$

Therefore,

$$\|(f * g)h\|_1 \leq \|g\|_1^{\frac{1}{p}} \|f\|_p \|g\|_1^{\frac{1}{p'}} \|h\|_{p'} = \|g\|_1 \|f\|_p \|h\|_{p'}.$$

Inequality (4.9) now follows from the duality between $L^p(B, \lambda)$ and $L^{p'}(B, \lambda)$. Inequality (4.10) is an easy consequence of Hölder's inequality and the invariance of the measure λ. $\square$

As a consequence of Lemma 4.4, if $g \in L^1_{loc}(B)$ is radial and $f \in L^p_{loc}(B)$ then $f * g$ is defined a.e. on B. There is one additional property of the above convolution which will be needed.

Lemma 4.5. ([Ul2]) *If f, χ, $h \in L^1(B, \lambda)$ and χ is radial, then $(f * \chi) * h = f * (\chi * h)$.*

Proof. Suppose $a, w \in B$. Since $(\varphi_{\varphi_a(w)} \circ \varphi_a \circ \varphi_w)(0) = 0$, by Cartan's theorem

$$\varphi_{\varphi_a(w)} \circ \varphi_a \circ \varphi_w \in \mathcal{U}.$$

Thus $\varphi_{\varphi_a(w)} = U \circ \varphi_w \circ \varphi_a$ for some $U \in \mathcal{U}$, and consequently $|\varphi_{\varphi_a(w)}(z)| = |\varphi_w(\varphi_a(z))|$. By the invariance of λ and the fact that χ is radial,

$$\int_B \chi(z) h(\varphi_{\varphi_a(w)}(z)) \, d\lambda(z) = \int_B \chi(\varphi_{\varphi_a(w)}(z) h(z) \, d\lambda(z)$$

$$= \int_B \chi(\varphi_w(\varphi_a(z))) h(z) \, d\lambda(z)$$

$$= \int_B \chi(\varphi_w(z)) h(\varphi_a(z)) \, d\lambda(z).$$

Therefore,

$$((f * \chi) * h)(a) = \int_B \int_B f(w) \chi(\varphi_z(w)) h(\varphi_a(z)) \, d\lambda(w) \, d\lambda(z)$$

$$= \int_B \int_B f(w) \chi(z) h(\varphi_{\varphi_a(w)}(z)) \, d\lambda(z) \, d\lambda(w)$$

$$= \int_B f(w)(\chi * h)(\varphi_a(w)) \, d\lambda(w) = (f * (\chi * h))(a). \quad \square$$

We now express identity (4.5) in terms of the above convolution. Define Ω_r by

$$\Omega_r(z) = \begin{cases} \dfrac{1}{\lambda(B(0,r))}, & |z| \leq r, \\[2mm] 0, & |z| > r. \end{cases}$$

Then by (4.5), if $f \in C^2(B)$,

$$(4.11) \qquad (\widetilde{\Delta} f)(a) = \lim_{r \to 0} \frac{4}{r^2} \left[(f * \Omega_r)(a) - f(a) \right].$$

If f has compact support, then the convergence is uniform on B.

4.3. The Riesz Measure.

Let $\{r_k\}$ be a decreasing sequence with $r_k \to 0$ and $k \to \infty$, and for each k let χ_k be a nonnegative C^∞ radial function on B with support contained in $\{z : r_{k+1} < |z| < r_k\}$ satisfying

$$\int_B \chi_k(z) \, d\lambda(z) = 1.$$

The family $\{\chi_k\}_{k=1}^\infty$ is a C^∞ **approximate identity** for $L^1(B, \lambda)$. One immediate consequence is the following:

Lemma 4.6. *Let* $\{\chi_k\}$ *be defined as above. Then*

$$\lim_{k\to\infty} (h * \chi_k) = h,$$

uniformly on B *if* $h \in C_c(B)$, *and locally in* L^p *if* $h \in L^p_{loc}(B)$.

For $\mathcal{M}$-subharmonic functions we obtain the following stronger result ([U12]) .

Proposition 4.7. *Let* $\{\chi_k\}$ *be a* C^∞ *approximate identity as above. If* f *is* $\mathcal{M}$-*subharmonic on* B, *then* $\{f * \chi_k\}_{k=1}^\infty$ *is a nonincreasing sequence of* C^∞ $\mathcal{M}$-*subharmonic functions satisfying*

$$(4.12) \qquad (f * \chi_k)(z) \geq f(z) \quad \text{and} \quad \lim_{k\to\infty} (f * \chi_k)(z) = f(z)$$

for all $z \in B$. *Furthermore, if* f *is* $\mathcal{M}$-*harmonic on* B, *then equality holds in* (4.12) *for all* $z \in B$.

Proof. Since χ_k is C^∞, $f * \chi_k$ is also C^∞. By integration in polar coordinates,

$$\begin{aligned}(f * \chi_k)(z) &= 2n \int_0^1 \frac{r^{2n-1}}{(1-r^2)^{n+1}} \chi_k(r) \left[\int_S f(\varphi_z(rt))\, d\sigma(t) \right] dr \\ &\geq f(z) \int_B \chi_k\, d\lambda = f(z),\end{aligned}$$

with equality if f is $\mathcal{M}$-harmonic.

Fix $z \in B$, and let $\alpha > f(z)$. Since f is upper semicontinuous there exists $r > 0$ such that $f(w) < \alpha$ for all $w \in E(z,r)$. Hence if $r_k < r$,

$$(f * \chi_k)(z) = \int_B f(w) \chi_k(\varphi_z(w))\, d\lambda(w) \leq \alpha \int_B \chi_k\, d\lambda = \alpha.$$

Thus $\limsup_{k\to\infty} (f * \chi_k)(z) \leq f(z)$, which when combined with the above proves (4.12).

To show that $f * \chi_k$ is $\mathcal{M}$-subharmonic we use (4.11). Since χ_k and Ω_r are both radial, by Lemma 4.3

$$(f * \chi_k) * \Omega_r = (f * \Omega_r) * \chi_k.$$

But as above, by integration in polar coordinates $(f * \Omega_r) \geq f$. Therefore

$$(f * \chi_k) * \Omega_r = (f * \Omega_r) * \chi_k \geq f * \chi_k.$$

Thus by (4.11) $\widetilde{\Delta}(f * \chi_k) \geq 0$. Therefore $f * \chi_k$ is $\mathcal{M}$-subharmonic.

Finally it remains to be shown that the sequence $\{f * \chi_k\}$ is nonincreasing. Since $(f * \chi_k)(a) = ((f \circ \varphi_a) * \chi_k)(0)$, it suffice to prove the result for $a = 0$. For the proof, we will need that

$$\int_S f(r_1 t) \, d\sigma(t) \leq \int_S f(r_2 t) \, d\sigma(t)$$

whenever $r_1 < r_2$, a fact which will be established in the next chapter (Proposition 5.11). Suppose $m > k$. Since the support of χ_k is contained in $\{r_{k+1} < |z| < r_k\}$ and $r_{k+1} \geq r_m$, we obtain

$$(f * \chi_k)(0) = 2n \int_{r_{k+1}}^{r_k} r^{2n-1} (1 - r^2)^{-n-1} \chi_k(r) \int_S f(rt) \, d\sigma(t) \, dr$$

$$\geq \int_S f(r_m t) \, d\sigma(t) \int_B \chi_k \, d\lambda$$

$$= \int_S f(r_m t) \, d\sigma(t) \int_B \chi_m \, d\lambda$$

$$\geq 2n \int_{r_{m+1}}^{r_m} r^{2n-1} (1 - r^2)^{-n-1} \chi_m(r) \int_S f(rt) \, d\sigma(t) \, dr$$

$$= (f * \chi_m)(0). \quad \square$$

Remark. One consequence of Proposition 4.7 is that if h is $\mathcal{M}$-harmonic on B, then h is C^∞ on B. This follows from that fact that since χ_k is C^∞, so is $h * \chi_k$. But by the proposition, $h * \chi_k = h$.

We are now ready to state and prove the following distributional characterization of $\mathcal{M}$-subharmonic and $\mathcal{M}$-harmonic functions on B.

Theorem 4.8. If f is $\mathcal{M}$-subharmonic on B, then

$$(4.13) \qquad \int_B f(z) \widetilde{\Delta} \psi(z) \, d\lambda(z) \geq 0$$

for all $\psi \in C_c^2(B)$ with $\psi \geq 0$. Conversely, if $f \in L^1_{loc}(B)$ is such that (4.13) holds for all $\psi \in C_c^2(B)$ with $\psi \geq 0$, then there exists an $\mathcal{M}$-subharmonic function F on B such that $F = f$ a.e. on B.

Proof. Let $\{\chi_k\}$ be a C^∞ approximate identity as defined above. Suppose f is $\mathcal{M}$-subharmonic on B. Set $f_k = f * \chi_k$. Then by Proposition 4.7 $\{f_k\}$ is a nonincreasing sequence of C^∞ $\mathcal{M}$-subharmonic functions of B which converges to f everywhere on B. Thus by Green's identity (3.3) and the monotone convergence theorem,

$$\int_B f \, \widetilde{\Delta} \psi \, d\lambda = \lim_{k \to \infty} \int_B f_k \, \widetilde{\Delta} \psi \, d\lambda$$

$$= \lim_{k \to \infty} \int_B \psi \, \widetilde{\Delta} f_k \, d\lambda \geq 0$$

for all $\psi \in C_c^2(B)$ with $\psi \geq 0$. This proves (4.13).

Conversely, suppose $f \in L_{loc}^1(B)$ satisfies (4.13). Let f_k be defined as above. In the notation of convolutions, hypothesis (4.13) is just

$$(f * \widetilde{\Delta}\psi)(0) \geq 0$$

for all $\psi \in C_c^2(B)$ with $\psi \geq 0$. It follows from the definition that

$$\widetilde{\Delta}f_k(z) = \widetilde{\Delta}(f * \chi_k)(z) = (f * \widetilde{\Delta}(\chi_k \circ \varphi_z))(0).$$

Thus since $\chi_k \circ \varphi_z$ is C^∞ with compact support, $\widetilde{\Delta}f_k(z) \geq 0$, and thus is $\mathcal{M}$-subharmonic on B.

We now show that the sequence $\{f_k\}$ is nonincreasing. Suppose $k > m$ and j is arbitrary. Since f_j is $\mathcal{M}$-subharmonic, we have

$$\begin{aligned} f_m * \chi_j = (f * \chi_m) * \chi_j &= (f * \chi_j) * \chi_m \\ &\geq (f * \chi_j) * \chi_k = f_k * \chi_j. \end{aligned}$$

Since f_m, f_k are continuous, $\lim_{j\to\infty} f_m * \chi_j = f_m$, with a similar result for f_k. Thus $\{f_k\}$ is nonincreasing. Define

$$F(z) = \lim_{k \to \infty} f_k(z)$$

which exists everywhere on B. As a consequence of (4.1), F is either $\mathcal{M}$-subharmonic on B, or $F \equiv -\infty$. But by Lemma 4.6, $\{f_k\}$ converges to f locally in L^1, and thus $F = f$ a.e. on B, which proves the result. $\square$

Corollary 4.9. *If h is $\mathcal{M}$-harmonic on B, then*

$$(4.14) \qquad \int_B h(z)\widetilde{\Delta}\psi(z)\, d\lambda(z) = 0$$

for all $\psi \in C_c^2(B)$. Conversely, if $h \in L_{loc}^1(B)$ is such that (4.14) holds for all $\psi \in C_c^2(B)$, then there exists an $\mathcal{M}$-harmonic function H on B such that $H = h$ a.e. on B.

Theorem 4.10. *If f is $\mathcal{M}$-subharmonic on B, then there exists a unique regular Borel measure μ_f on B such that*

$$(4.15) \qquad \int_B \psi\, d\mu_f = \int_B f\widetilde{\Delta}\psi\, d\lambda$$

for all $\psi \in C_c^2(B)$.

Definition. *If f is $\mathcal{M}$-subharmonic on B, the unique regular Borel measure μ_f satisfying (4.15) is called the* **Riesz measure** *of f.*

Proof. Let f be $\mathcal{M}$-subharmonic on B. By (4.13),

$$L(\psi) = \int_B f\,\widetilde{\Delta}\psi\,d\lambda$$

defines a nonnegative linear functional on $C_c^\infty(B)$. We extend L to $C_c(B)$ as follows. Let $\psi \in C_c(B)$. Choose a sequence $\{\psi_k\} \subset C_c^\infty(B)$ such that $\psi_k \to \psi$ uniformly on B. Choose a compact subset K of B such that the support of ψ and ψ_k, $k = 1, 2...$ are contained in K. Let V be a relatively compact open subset of B such that $K \subset V$, and let $h \in C_c^\infty(B)$, $0 \le h \le 1$, be such that $h \equiv 1$ on K and support of h is contained in V. Let

$$\epsilon_{k,m} = \sup_{x \in K} |\psi_k(x) - \psi_m(x)|.$$

Then for all $x \in B$,

$$-\epsilon_{k,m}h(x) \le \psi_k(x) - \psi_m(x) \le \epsilon_{k,m}h(x).$$

Thus since L is positive,

$$|L(\psi_k) - L(\psi_m)| \le \epsilon_{k,m}\,L(h).$$

Therefore $\{L(\psi_k)\}$ is Cauchy. Define

$$L(\psi) = \lim_{k \to \infty} L(\psi_k).$$

It is easy to show that $L(\psi)$ is independent of the choice of $\{\psi_k\}$, and thus defines a nonnegative linear functional on $C_c(B)$. The result now follows by the Riesz representation theorem for nonnegative linear functionals on $C_c(B)$. $\square$

4.4. Remarks.

(1) For a bounded symmetric domain D, there is a characterization analogous to (4.1) for strongly harmonic functions on D. Let G be the connected component of the identity of $Aut(D)$ and let

$$K = \{g \in G : g \cdot o = o\},$$

where $o \in D$ is fixed. K is called the isotropy subgroup of G at o, which is compact. By a result of Godement [Go], a continuous real valued function f on D is strongly harmonic if and only if for all $g \in G$,

$$(4.16) \qquad\qquad f(g \cdot o) = \int_K f(gk \cdot z)\,dk$$

for all z in a neighborhood of the point o. In the above, dk denotes the normalized Haar measure on K. However, except for the case where the class of strongly and weakly harmonic functions coincide (e.g. the ball or rank 1 symmetric spaces), no such characterization is available for weakly harmonic functions.

Using (4.16) one can define strongly subharmonic functions as follows: an upper-semicontinuous function $f : D \to [-\infty, \infty)$ is strongly subharmonic on D if for all $g \in G$,

$$(4.17) \qquad f(g \cdot o) \leq \int_K f(gk \cdot z) \, dk$$

for all z in a neighborhood of $o \in D$.

We illustrate this for the polydisc U^n. The automorphism group $Aut(U^n)$ is characterized as follows ([Kr1, Proposition 10.1.3]): Let $\psi \in Aut(U^n)$. Then there exist $a_1, ..., a_n \in U$, $\theta_i \in [0, 2\pi)$, and a permutation σ of $\{1, ..., n\}$ such that

$$\psi(z) = \left(e^{i\theta_1} \frac{a_1 - z_{\sigma(1)}}{1 - \bar{a}_1 z_{\sigma(1)}}, \, ..., \, e^{i\theta_n} \frac{a_n - z_{\sigma(n)}}{1 - \bar{a}_n z_{\sigma(n)}} \right).$$

The connected component G of the identity is the subgroup given by all automorphisms of the form

$$\psi(z) = \left(e^{i\theta_1} \varphi_{a_1}(z_1), \, ..., \, e^{i\theta_n} \varphi_{a_n}(z_n) \right),$$

where φ_{a_i} is the usual Moebius transformation of U. In this case, the subgroup K of G fixing the origin is just T^n. If f is n-subharmonic on U^n, and

$$\varphi_a(z) = (\varphi_{a_1}(z_1), \, ..., \, \varphi_{a_n}(z_n)) \in G,$$

then $f \circ \varphi_a$ is also n-subharmonic for all $a = (a_1, ..., a_n) \in U^n$. Thus for all $a \in U^n$,

$$(4.18) \qquad f(a) \leq \int_{T^n} f(\varphi_{a_1}(r_1 t_1), ..., \varphi_{a_n}(r_n t_n)) \, d\sigma(t),$$

for all r_j, $j = 1, ..., n$ sufficiently small. In the above, σ denotes normalized Lebesgue measure on T^n. Inequality (4.18) is just (4.17) for the specific case of the polydisc. It is clear that if f satisfies (4.18), then f is n-subharmonic on B.

For C^2 functions, weak subharmonicity can be defined in terms of the Laplace-Beltrami operator $\widetilde{\Delta}_D$. I am not aware of any significant studies concerning either strongly or weakly subharmonic functions on any domains other than the ball or the polydisc. Even in the case of the polydisc, virtually nothing is known about the class of weakly subharmonic functions.

(2) Our second remark concerns an interesting question which has only recently been solved. Suppose f is $\mathcal{M}$-harmonic on B. Since $f \circ \psi$ is also $\mathcal{M}$-harmonic, by (4.1) and integration in polar coordinates,

$$(4.19) \qquad\qquad f(\psi(0)) = \int_B f(\psi(w))\, d\nu(w)$$

for every $\psi \in Aut(B)$. One is therefore naturally led to ask the following:

Question. *If $f \in L^1(B,\nu)$ and satisfies (4.19) for every $\psi \in Aut(B)$, is f $\mathcal{M}$-harmonic?*

It has been known for sometime that the answer is yes for all n if $f \in C(\overline{B})$ [NR]. The recent rather surprising result of P. Ahern, M. Flores and W. Rudin [NFR] answers the question in the affirmative if $n \leq 11$, and in the negative when $n \geq 12$.

5.
Poisson-Szegö Integrals

In this chapter we derive a Poisson integral formula for $\mathcal{M}$-harmonic functions on B, and consider some of the properties of the Poisson kernel. One of the difficulties encountered in dealing with $\mathcal{M}$-harmonic functions as opposed to holomorphic, pluriharmonic, or harmonic (as in $\mathbb{R}^n$) is that the slice function f_r defined by $f_r(z) = f(rz)$ is in general not $\mathcal{M}$-harmonic if f is. In fact, if f is $\mathcal{M}$-harmonic and there exists a r, $0 < r < 1$, such that f_r is also $\mathcal{M}$-harmonic on B, then f is pluriharmonic on B ([Ru3, Theorem 4.4.10]).

5.1. The Poisson-Szegö Kernel.

We begin by first deriving the Cauchy-Szegö integral formula for holomorphic functions on B. Let $A(B)$ denote the class of holomorphic functions f on B which are continuous on $\overline{B}$.

Proposition 5.1. For $f \in A(B)$,

$$(5.1) \qquad f(z) = \int_S \frac{f(t)}{(1 - \langle z, t \rangle)^n} \, d\sigma(t), \qquad z \in B.$$

Proof. Let

$$f(z) = \sum_\alpha a_\alpha z^\alpha$$

be the Taylor series expansion of f about 0, which since B is a Reinhardt domain converges uniformly on compact subsets of B. Thus for $0 < r < 1$,

$$\int_S \frac{f(rt)}{(1 - \langle z, t \rangle)^n} \, d\sigma(t) = \sum_{k=0}^\infty r^k \sum_{|\alpha|=k} a_\alpha \int_S \frac{t^\alpha}{(1 - \langle z, t \rangle)^n} \, d\sigma(t).$$

By the binomial expansion (1.11),

$$(1 - \langle z, t \rangle)^{-n} = \sum_{k=0}^\infty \frac{\Gamma(k+n)}{\Gamma(n)\Gamma(k+1)} \langle z, t \rangle^k.$$

Thus

$$\int_S \frac{t^\alpha}{(1 - \langle z, t \rangle)^n} \, d\sigma(t) = \sum_{k=0}^\infty \frac{\Gamma(k+n)}{\Gamma(n)\Gamma(k+1)} \int_S t^\alpha \langle z, t \rangle^k \, d\sigma(t)$$

$$= \frac{\Gamma(|\alpha|+n)}{\Gamma(n)\Gamma(|\alpha|+1)} \int_S t^\alpha \langle z, t \rangle^{|\alpha|} \, d\sigma(t).$$

The last equality follows since $\int t^\alpha \langle z, t \rangle^k \, d\sigma(t) = 0$ for $k \neq |\alpha|$. Finally, since

$$\langle z, t \rangle^{|\alpha|} = \sum_{|\beta| = |\alpha|} \frac{\Gamma(|\alpha| + 1)}{\beta!} z^\beta \, \bar{t}^\beta,$$

by orthogonality and identity (2.9),

$$\int_S t^\alpha \langle z, t \rangle^{|\alpha|} \, d\sigma(t) = z^\alpha \frac{\Gamma(|\alpha| + 1)}{\alpha!} \frac{\Gamma(n)\alpha!}{\Gamma(n + |\alpha|)}.$$

Therefore

$$\int_S \frac{t^\alpha}{(1 - \langle z, t \rangle)^n} \, d\sigma(t) = z^\alpha,$$

from which the result now follows by letting $r \to 1$. $\quad\square$

The kernel

$$S(z, w) = \frac{1}{(1 - \langle z, w \rangle)^n} \qquad (z, w) \in B \times \overline{B},$$

is called the **Szegö** or **Cauchy-Szegö** kernel of B. When $n = 1$, $S(z, t)$, $z \in U$, $t \in T$ is the Cauchy-kernel of U. The integral formula (5.1) seems to have appeared first in L. K. Hua's book [Hu] in 1958 for bounded symmetric domains in $\mathbb{C}^n$, and subsequently in 1964 in a paper by S. G. Gindiken ([Gi]) for Siegel domains of type II. For domains of type D_α it has also been computed by A. Bonami and N. Lohoué in [BL]. The Szegö kernel can be shown to be the reproducing kernel of the classical Hardy space $H^2(B)$, which will be defined shortly.

Definition. *The* **Poisson-Szegö** *or* **Poisson** *kernel of B is defined by*

$$(5.2) \qquad \mathcal{P}(z, t) = \frac{|S(z, t)|^2}{S(z, z)} = \frac{(1 - |z|^2)^n}{|1 - \langle z, t \rangle|^{2n}}, \qquad z \in B, \, t \in S.$$

The Poisson-Szegö kernel can be defined analogously for any domain which has a Szegö kernel. It was shown by A. Korany in [Ko1] that the Poisson-Szegö kernel for any bounded symmetric domain is strongly harmonic. We will establish this fact for B shortly by direct computation. This result however is not true in general. For the domains D_α, $\alpha \neq 1$, the Poisson-Szegö kernel is given by

$$\mathcal{P}(z, t) = \frac{(\alpha + 1)}{\pi} \left(\frac{|1 - z_2 \bar{t}_2|^2}{1 - |z_2|^2} \right)^{\alpha - 1} \frac{\left[(1 - |z_2|^2)^\alpha - |z_1|^2 \right]^2}{|(1 - z_2 \bar{t}_2)^\alpha - z_1 \bar{t}_1|^4},$$

and it can be shown that it is not annihilated by the Laplace-Beltrami operator for D_α ([Ha]).

One immediate consequence of Proposition 5.1 is the following:

Corollary 5.2. If $f \in A(B)$, then

$$(5.3) \qquad f(z) = \int_S \mathcal{P}(z,t) f(t) \, d\sigma(t), \qquad z \in B.$$

Proof. Fix $z \in B$, and set

$$g(w) = \frac{S(w,z)}{S(z,z)} f(w), \qquad w \in B.$$

Then $g \in A(B)$ and $g(z) = f(z)$. Thus

$$f(z) = \int_S f(t) \frac{S(t,z)}{S(z,z)} S(z,t) \, d\sigma(t) = \int_S \mathcal{P}(z,t) f(t) \, d\sigma(t). \quad \square$$

The Szegö kernel has the advantage that given any $f \in C(S)$, the Szegö integral of f is holomorphic on B. One of the difficulties encountered with the Szegö kernel is that it is not positive and that

$$\int_S |S(z,t)| \, d\sigma(t) \to +\infty$$

as z approaches any boundary point of B. The advantage of the Poisson-Szegö kernel is that it is nonnegative, and as we will shortly see, forms an approximate identity.

Before proceeding further, we will compare the singularities of the classical Poisson kernel and the Poisson-Szegö kernel. The Poisson kernel $P(z,t)$ for the Laplacian Δ on the ball is given by

$$(5.4) \qquad P(z,t) = \frac{(1 - |z|^2)}{|z - t|^{2n}}, \qquad z \in B, t \in S.$$

The singularity of $P(z,t)$ on the boundary is $|z-t|^{2n}$, whereas the singularity of $\mathcal{P}(z,t)$ on the boundary is $|1 - \langle z,t \rangle|^{2n}$.

The natural geometry on S associated with the singularity $|z - t|$ are the balls given by

$$B(t,r) = \{\zeta \in S : |t - \zeta| < r\}.$$

These balls are **isotropic**; they are the same in all directions. The natural geometry in S associated with the singularity $|1 - \langle z,t \rangle|$ are the balls

$$(5.5) \qquad Q(t,r) = \{\zeta \in S : |1 - \langle t,\zeta \rangle| < r\}.$$

To understand the geometry of the balls $Q(t,r)$, $t \in S$, let $t = e_1$, and write $\zeta = (\zeta_1, \zeta')$. Then

$$Q(t,r) = \{(\zeta_1, \zeta') \in S : |1 - \zeta_1| < r\}.$$

Since $\zeta \in S$, we have

$$|\zeta'|^2 = 1 - |\zeta_1|^2 = (1 - |\zeta_1|)(1 + |\zeta_1|)$$
$$\leq 2\,(1 - |\zeta_1|) \leq 2\,|1 - \zeta_1| < 2r.$$

Therefore, $Q(e_1, r)$ has radius r in the e_1 direction, and radius of order $\sqrt{r}$ in the orthogonal direction. The balls $Q(t, r)$ are therefore **nonisotropic**.

Remark: Although we will not need it at this point, we will subsequently show that the function

(5.6) $d(z, w) = |1 - \langle z, w \rangle|^{1/2}$

satisfies the triangle inequality on $\overline{B}$. Although it is not a metric on $\overline{B}$, it is a metric on S, which is generally referred to as the **nonisotropic** metric.

Lemma 5.3.
   (a) $\mathcal{P}(z, t) \geq 0$.
   (b) $\mathcal{P}(rt, \zeta) = \mathcal{P}(r\zeta, t)$ for all $t, \zeta \in S$.
   (c) $\int_S \mathcal{P}(z, t)\, d\sigma(t) = 1$, and
   (d) for $\zeta \in S$, $\epsilon > 0$,

$$\lim_{z \to \zeta} \int_{S \sim Q(\zeta, \epsilon)} \mathcal{P}(z, t)\, d\sigma(t) = 0.$$

Proof. The proofs of (a) and (b) follow from the definition. The proof of (c) follows by taking $f \equiv 1$ in (5.3).
   (d) If $t \in S \sim Q(\zeta, \epsilon)$,

$$\epsilon \leq |1 - \langle \zeta, t \rangle| \leq |1 - \langle z, t \rangle| + |\langle z - \zeta, t \rangle|$$
$$\leq |1 - \langle z, t \rangle| + |z - \zeta|.$$

Thus if $|z - \zeta| < \epsilon/2$,

$$|1 - \langle z, t \rangle| \geq \frac{\epsilon}{2} \qquad \text{for all} \quad t \in S \sim Q(\zeta, \epsilon).$$

Therefore, for all $z \in B$ such that $|z - \zeta| < \epsilon/2$,

$$\int_{S \sim Q(\zeta, \epsilon)} \mathcal{P}(z, t)\, d\sigma(t) \leq \frac{4^n}{\epsilon^{2n}}\,(1 - |z|^2)^n,$$

from which the result follows. $\quad\square$

Proposition 5.4. *For fixed* $t \in S$, $\mathcal{P}(z,t)$ *is* $\mathcal{M}$*-harmonic on* B. *Furthermore, for* $\alpha > 0$,

$$(5.7) \qquad \tilde{\Delta}\mathcal{P}^\alpha(z,t) = \frac{4n^2}{n+1}\,\alpha(\alpha-1)\mathcal{P}^\alpha(z,t).$$

Proof. To prove that $\mathcal{P}(z,t)$ is $\mathcal{M}$-harmonic on B, we will show that $\mathcal{P}$ satisfies the invariant mean value equality (4.1). By identity (1.15), for a, $z \in B$, $\zeta \in S$,

$$S(\varphi_a(z),\varphi_a(\zeta)) = \frac{S(a,a)S(z,\zeta)}{S(z,a)S(a,\zeta)}.$$

Therefore

$$(5.8) \qquad \mathcal{P}(\varphi_a(z),\varphi_a(\zeta)) = \frac{\mathcal{P}(z,\zeta)}{\mathcal{P}(a,\zeta)}.$$

As a consequence, for $0 < r < 1$ and $\eta \in S$,

$$\int_S \mathcal{P}(\varphi_a(rt),\eta)\,d\sigma(t) = \frac{1}{\mathcal{P}(a,\varphi_a(\eta))}\int_S \mathcal{P}(rt,\varphi_a(\eta))\,d\sigma(t)$$

$$= \mathcal{P}(\varphi_a(0),\eta)\int_S \mathcal{P}(r\varphi_a(\eta),t)\,d\sigma(t)$$

$$= \mathcal{P}(a,\eta).$$

In the above we have used (b) and (c) of Lemma 5.3.

To prove (5.7), fix $t \in S$, and write $\mathcal{P}_t(z) = \mathcal{P}(z,t)$. By (5.8),

$$\mathcal{P}(\varphi_a(z),t) = \mathcal{P}(z,\zeta)\mathcal{P}(a,t),$$

where $\zeta = \varphi_a(t)$. Thus by (3.8),

$$\left(\tilde{\Delta}\mathcal{P}_t^\alpha\right)(a) = \frac{1}{n+1}\Delta\left(\mathcal{P}_t^\alpha \circ \varphi_a\right)(0)$$

$$= \frac{1}{n+1}\mathcal{P}^\alpha(a,t)\Delta\mathcal{P}_\zeta^\alpha(0).$$

Since $\mathcal{P}$ is $\mathcal{M}$-harmonic,

$$\Delta\mathcal{P}_\zeta^\alpha(0) = 4\sum_{j=1}^n \frac{\partial^2 \mathcal{P}_\zeta^\alpha}{\partial z_j \partial \overline{z}_j}(0)$$

$$= 4\alpha(\alpha-1)\mathcal{P}^{\alpha-2}(0,\zeta)\sum_{j=1}^n \left|\frac{\partial \mathcal{P}_\zeta}{\partial z_j}(0)\right|^2.$$

But $\mathcal{P}(0,\zeta) = 1$ and $\dfrac{\partial \mathcal{P}_\zeta}{\partial z_j}(0) = n\overline{\zeta}_j$. Therefore

$$\Delta\mathcal{P}_\zeta^\alpha(0) = 4n^2\,\alpha(\alpha-1),$$

from which the result now follows. $\square$

5.2. The Dirichlet Problem for $\widetilde{\Delta}$.

We now turn to the solution of the Dirichlet problem for the invariant Laplacian $\widetilde{\Delta}$ on B.

Theorem 5.5. *If $f \in C(S)$, then the function*

$$(5.9) \qquad F(z) = \begin{cases} \displaystyle\int_S \mathcal{P}(z,t)f(t)\,d\sigma(t), & z \in B \\ f(z), & z \in S \end{cases}$$

is $\mathcal{M}$-harmonic on B and continuous on $\overline{B}$.

Proof. By Fubini's theorem, since $\mathcal{P}$ is $\mathcal{M}$-harmonic, the function F satisfies the invariant mean value equality (4.1) and thus is $\mathcal{M}$-harmonic on B. To prove continuity on $\overline{B}$, let $\zeta \in S$, and fix $\epsilon > 0$. Since f is continuous at ζ, there exists a $\delta > 0$ such that

$$|f(t) - f(\zeta)| < \epsilon \qquad \text{for all} \quad t \in Q(\zeta, \delta).$$

Therefore,

$$
\begin{aligned}
|F(z) - f(\zeta)| &= \left| \int_S \mathcal{P}(z,t)f(t)\,d\sigma(t) - f(\zeta) \right| \\
&\le \int_{Q(\zeta,\delta)} \mathcal{P}(z,t)|f(t) - f(\zeta)|\,d\sigma(t) \\
&\qquad + \int_{S \sim Q(\zeta,\delta)} \mathcal{P}(z,t)|f(t) - f(\zeta)|\,d\sigma(t) \\
&\le \epsilon + 2\|f\|_\infty \int_{S \sim Q(\zeta,\delta)} \mathcal{P}(z,t)\,d\sigma(t).
\end{aligned}
$$

Thus by Lemma 5.3, $\lim_{z \to \zeta} F(z) = f(\zeta)$, which proves the result. $\square$

There is a significant difference between invariant Poisson integrals on the ball in $\mathbb{C}^n$, and solutions of the classical Dirichlet problem for the Laplacian Δ. Since Δ is uniformly elliptic, if $f \in C^\infty(S)$ and F is the euclidean Poisson integral of f, then F is C^∞ on $\overline{B}$. The following example shows that this result fails very dramatically for invariant Poisson integrals.

Example. ([Kr2]). Let $n = 2$ and let f be defined by

$$f(t_1, t_2) = |t_1|^2.$$

Then $f \in C^\infty(S)$. We now compute the invariant Poisson integral F of f.

$$F(z) = (1 - |z|^2)^2 \int_S \frac{|t_1|^2}{|1 - \langle z, t \rangle|^4}\,d\sigma(t).$$

By the binomial expansion (1.11),

$$|1 - \langle z, t\rangle|^{-4} = (1 - \langle z, t\rangle)^{-2} (1 - \overline{\langle z, t\rangle})^{-2}$$

$$= \sum_{m,k=0}^{\infty} \frac{\Gamma(m+2)\Gamma(k+2)}{\Gamma(m+1)\Gamma(k+1)} \langle z, t\rangle^k \overline{\langle z, t\rangle}^m.$$

Therefore,

$$\int_S \frac{|t_1|^2}{|1 - \langle z, t\rangle|^4} \, d\sigma(t) = \sum_{m,k=0}^{\infty} (m+1)(k+1) \int_S |t_1|^2 \langle z, t\rangle^k \overline{\langle z, t\rangle}^m \, d\sigma(t)$$

$$= \sum_{m=0}^{\infty} (m+1)^2 \int_S |t_1|^2 |\langle z, t\rangle|^{2m} \, d\sigma(t).$$

The last identity follows from orthogonality. By the multinomial expansion and orthogonality,

$$\int_S |t_1|^2 |\langle z, t\rangle|^{2m} \, d\sigma(t) = \sum_{\substack{\alpha,\beta \\ |\alpha|=|\beta|=m}} \frac{(m!)^2}{\alpha!\beta!} z^\alpha \bar{z}^\beta \int_S |t_1|^2 \bar{t}^\alpha t^\beta \, d\sigma(t)$$

$$= \sum_{|\alpha|=m} \frac{(m!)^2}{(\alpha!)^2} |z^\alpha|^2 \int_S |t_1|^2 |t^\alpha|^2 \, d\sigma(t).$$

But by (2.9)

$$\int_S |t_1|^2 |t^\alpha|^2 \, d\sigma(t) = \int_S |t_1^{\alpha_1+1}|^2 |t_2^{\alpha_2}|^2 \, d\sigma(t) = \frac{(\alpha_1+1)!\alpha_2!}{(m+2)!}.$$

Therefore

$$\int_S |t_1|^2 |\langle z, t\rangle|^{2m} \, d\sigma(t) = \frac{1}{(m+2)(m+1)} \sum_{|\alpha|=m} \frac{m!(\alpha_1+1)}{\alpha_1!\alpha_2!} |z_1|^{2\alpha_1} |z_2|^{2\alpha_2}$$

$$= \frac{1}{(m+2)(m+1)} \sum_{k=0}^{m} \frac{m!(k+1)}{k!(m-k)!} |z_1|^{2k} |z_2|^{2(m-k)}.$$

Since

$$x\,(x+y)^m = \sum_{k=0}^{m} \frac{m!}{k!(m-k)!} x^{k+1} y^{m-k},$$

by taking the partial with respect to x we obtain

$$\sum_{k=0}^{m} \frac{m!(k+1)}{k!(m-k)!} x^k y^{m-k} = (x+y)^m + m\,x\,(x+y)^{m-1}.$$

Thus with $x = |z_1|^2$ and $y = |z_2|^2$,

$$\int_S |t_1|^2 |\langle z, t \rangle|^{2m} \, d\sigma(t) = \frac{1}{(m+2)(m+1)} \left[|z|^{2m} + m |z_1|^2 |z|^{2(m-1)} \right],$$

and as a consequence

$$F(z) = (1 - |z|^2)^2 \sum_{m=0}^{\infty} \frac{m+1}{m+2} \left[|z|^{2m} + m |z_1|^2 |z|^{2(m-1)} \right].$$

We now evaluate each of the sums. For convenience let $x = |z|^2$. The first term becomes

$$\sum_{m=0}^{\infty} \frac{m+1}{m+2} x^m = \sum_{m=0}^{\infty} x^m - \sum_{m=0}^{\infty} \frac{1}{m+2} x^m$$

$$= \frac{1}{(1-x)} - \frac{1}{x^2} \sum_{m=2}^{\infty} \frac{x^m}{m}$$

$$= \frac{1}{(1-x)} + \frac{1}{x^2} \left[x + \log(1-x) \right].$$

For the second term,

$$\sum_{m=0}^{\infty} \frac{m(m+1)}{m+2} x^{m-1} = \sum_{m=1}^{\infty} \frac{m(m+1)}{m+2} x^{m-1}$$

$$= \sum_{m=1}^{\infty} m \, x^{m-1} - \sum_{m=1}^{\infty} x^{m-1} + 2 \sum_{m=1}^{\infty} \frac{x^{m-1}}{m+2}$$

$$= \frac{1}{(1-x)^2} - \frac{1}{(1-x)} + \frac{2}{x^3} \sum_{m=3}^{\infty} \frac{x^m}{m}$$

$$= \frac{1}{(1-x)^2} - \frac{1}{(1-x)} + \frac{2}{x^3} \left[-\log(1-x) - x - \frac{x^2}{2} \right].$$

Therefore,

$$F(z) = (1 - |z|^2)(1 - |z_1|^2) + |z_1|^2$$
$$+ \frac{(1 - |z|^2)^2}{|z|^4} \left[|z|^2 + \log(1 - |z|^2) \right]$$
$$- \frac{2|z_1|^2 (1 - |z|^2)^2}{|z|^6} \left[\log(1 - |z|^2) + |z|^2 + \frac{|z|^4}{2} \right].$$

From the above we see that for $t \in S$,

$$\lim_{z \to t} F(z) = |t_1|^2.$$

Consider $F(re_1)$, $0 < r < 1$. After simplification,

$$F(re_1) = (1 - r^2)^2 + r^2 - \frac{(1 - r^2)^2(1 + r^2)}{r^2} - \frac{(1 - r^2)^2}{r^4} \log(1 - r^2).$$

Thus $F(r)$ is the sum of four terms. The first three are C^∞ at the boundary point $e_1 = (1, 0)$; the last term however is not C^2 (from the left) at 1.

The phenomenon described in this example was discovered by Garnett and Krantz in 1977 (unpublished) and independently by C. R. Graham. Using spherical harmonics, one can show that the Poisson-Szegö integral of a C^∞ function f on S is C^∞ in $\overline{B}$ if and only if f is the boundary function of a pluriharmonic function. ([Kr2]).

5.3. Poisson-Szegö Integrals.

For $f \in L^p(S)$, $1 \le p \le \infty$, set

$$(5.10) \qquad \mathcal{P}[f](z) = \int_S \mathcal{P}(z, t) f(t) \, d\sigma(t),$$

and if ν is a complex (or finite signed) measure on S, let

$$(5.11) \qquad \mathcal{P}[\nu](z) = \int_S \mathcal{P}(z, t) \, d\nu(t).$$

By Fubini's theorem the functions $\mathcal{P}[\nu]$ and $\mathcal{P}[f]$ satisfy the invariant mean value equality (4.1), and thus are $\mathcal{M}$-harmonic on B. Since $\mathcal{P}(z, t) d\sigma(t)$ is a probability measure on S, by Jensen's inequality, for $1 \le p < \infty$,

$$|\mathcal{P}[f](r\zeta)|^p \le \int_S \mathcal{P}(r\zeta, t) |f(t)|^p \, d\sigma(t).$$

Therefore,

$$(5.12) \qquad \sup_{0 < r < 1} \int_S |\mathcal{P}[f](r\zeta)|^p \, d\sigma(\zeta) \le \|f\|_p^p.$$

Similarly, when $p = \infty$, we have $\|\mathcal{P}[f]\|_\infty \le \|f\|_\infty$, and also

$$(5.13) \qquad \sup_{0 < r < 1} \int_S |\mathcal{P}[\nu](rt)| \, d\sigma(t) \le \|\nu\|,$$

where $\|\nu\|$ denotes the total variation of the complex measure ν. As a consequence of Lemma 5.3 and Theorem 5.5 we have the following:

Proposition 5.6.

(a) If $f \in C(S)$, then $\lim_{r \to 1} \mathcal{P}[f](rt) = f(t)$ uniformly on S.

(b) If $f \in L^p(S)$, $1 \le p < \infty$, then

$$\lim_{r \to 1} \int_S |\mathcal{P}[f](rt) - f(t)|^p \, d\sigma(t) = 0.$$

(c) If ν is a complex (or finite signed) measure on S, then for all $\psi \in C(S)$,

$$\lim_{r \to 1} \int_S \mathcal{P}[\nu](rt)\psi(t) \, d\sigma(t) = \int_S \psi(t) \, d\nu(t).$$

Proof. (a) and (b) are immediate consequences of Lemma 5.3 and Theorem 5.5. For the proof of (c), if $\psi \in C(S)$, then by Fubini's theorem,

$$\int_S \mathcal{P}[\nu](rt)\psi(t) \, d\sigma(t) = \int_S \int_S \mathcal{P}(rt, \zeta) \, d\nu(\zeta) \, \psi(t) \, d\sigma(t)$$

$$= \int_S \int_S \mathcal{P}(r\zeta, t)\psi(t) \, d\sigma(t) \, d\nu(\zeta) = \int_S \mathcal{P}[\psi](r\zeta) \, d\nu(\zeta).$$

But $\mathcal{P}[\psi](r\zeta) \to \psi(\zeta)$ uniformly on S, from which the result follows. □

Corollary 5.7. If ν is a complex (or finite signed) measure on S for which $\mathcal{P}[\nu](z) = 0$ for all $z \in B$, then $\nu = 0$.

Proof. Part (c) of the Proposition implies that $\int_S \psi \, d\nu = 0$ for all $\psi \in C(S)$, which proves the result. □

For a function F defined on B, and $0 < r < 1$, let $F_r(z) = F(rz)$. Set

$$M_p(F, r) = \|F_r\|_p = \left[\int_S |F(rt)|^p \, d\sigma(t) \right]^{1/p}, \qquad 0 < p < \infty, \quad \text{and}$$

$$M_\infty(F, r) = \|F_r\|_\infty = \sup\{|F(rt)| : t \in S\}.$$

For $0 < p \le \infty$, let

$$\mathcal{H}^p(B) = \{F : F \text{ is } \mathcal{M}\text{-harmonic on B, and } \sup_{0 < r < 1} M_p(F, r) < \infty\}.$$

The Hardy spaces $H^p(B)$ of holomorphic functions on B are defined similarly. Also, one can consider the spaces $\mathcal{P}h^p$ of pluriharmonic functions. If $F(z) = \mathcal{P}[f](z)$, where $f \in L^p(S)$, $1 \le p \le \infty$, then by (5.12) $F \in \mathcal{H}^p(B)$. We now prove the converse.

Theorem 5.8. *Let F be a continuous $\mathcal{M}$-subharmonic function on B satisfying*

$$(5.14) \qquad \sup_{0<r<1} M_p(F,r) = \|F\|_p < \infty$$

for some p, $1 \leq p \leq \infty$.

  (a) If $1 < p \leq \infty$, then there exists $f \in L^p(S)$ such that

$$(5.15) \qquad F(z) \leq \mathcal{P}[f](z), \qquad z \in B.$$

  (b) If $p = 1$, then there exists a regular Borel measure ν on B such that

$$(5.16) \qquad F(z) \leq \mathcal{P}[\nu](z), \qquad z \in B.$$

  (c) If F is $\mathcal{M}$-harmonic on B and satisfies (5.14) for some p, $1 \leq p \leq \infty$, then equality holds in (5.15) (or (5.16) if $p = 1$).

Remark: For harmonic functions, the case $p = \infty$ was originally proved by Furstenberg ([Fu1]) for symmetric spaces of noncompact type. The L^p statements were proved by Koranyi in [Ko2] for the ball, and by the author ([Sto2]) for symmetric spaces of noncompact type. The following is a variation of an elementary proof using an equicontinuity argument which for harmonic functions is due to David Ullrich [Ul1]. (see also [Ru3, Theorem 4.3.3])

Proof. Let $h : \mathcal{U} \to [0,\infty)$ be a continuous function satisfying

$$\int_{\mathcal{U}} h(U)\,dU = 1.$$

where dU is the normalized Haar measure on $\mathcal{U}$. Let F be a continuous subharmonic function on B satisfying (5.14) for some p, $1 \leq p \leq \infty$. Define G on B by

$$(5.17) \qquad G(z) = \int_{\mathcal{U}} F(Uz)h(U)\,dU.$$

By Fubini's theorem, G satisfies (4.1) and thus is $\mathcal{M}$-subharmonic on B. If F is $\mathcal{M}$-harmonic, so is G. Since $\|F\|_p < \infty$, by Hölder's inequality applied to (5.17), for $z = r\zeta$, $\zeta \in S$,

$$(5.18) \qquad |G(z)| \leq \left[\int_{\mathcal{U}} |F(Ur\zeta)|^p \, dU \right]^{1/p} \|h\|_q \leq \|F\|_p \, \|h\|_q,$$

where $1/p + 1/q = 1$. In the above we have used the fact that

$$\int_{\mathcal{U}} |F(Ur\zeta)|^p \, dU = \int_S |F(rt)|^p \, d\sigma(t)$$

which follows by (1.9). Also by the continuous version of Minkowski's inequality, for $1 \leq p < \infty$,

$$\left[\int_S |G(r\zeta)|^p \, d\sigma(\zeta) \right]^{1/p} \leq \int_{\mathcal{U}} h(U) \left[\int_S |F(Ur\zeta)|^p \, d\sigma(\zeta) \right]^{1/p} dU$$
$$= \|F_r\|_p \leq \|F\|_p.$$

Since the result is also true for $p = \infty$,

(5.19) $$\qquad M_p(G, r) \leq \|F\|_p, \qquad (0 < r < 1),$$

for all p, $1 \leq p \leq \infty$.

We now show that $\{G_r : 0 < r < 1\}$ is an equicontinuous family of functions on S. Let $\epsilon > 0$ be fixed. Since h is continuous on $\mathcal{U}$, there exists a neighborhood $\mathcal{V}$ of the identity I such that

$$|h(U) - h(UV^{-1})| < \epsilon$$

for all $U \in \mathcal{U}$, $V \in \mathcal{V}$. Since

$$G_r(V\zeta) = \int_{\mathcal{U}} F(rUV\zeta)h(U) \, dU = \int_{\mathcal{U}} F(rU\zeta)h(UV^{-1}) \, dU,$$

we obtain

$$|G_r(\zeta) - G_r(V\zeta)| \leq \int_{\mathcal{U}} F(rU\zeta)|h(U) - h(UV^{-1})| \, dU$$
$$\leq \|F\|_1 \epsilon \leq \|F\|_p \epsilon$$

for all $V \in \mathcal{V}$. The mapping $U \to U\zeta$ is a one-to-one continuous mapping of $\mathcal{U}$ onto S. Thus there exists a $\delta > 0$ such that $|\zeta - \eta| < \delta$ implies $\eta = V\zeta$ for some $V \in \mathcal{V}$. Therefore

$$|G_r(\zeta) - G_r(\eta)| \leq \|F\|_p \epsilon$$

for all $\zeta, \eta \in S$ with $|\zeta - \eta| < \delta$. Thus the family $\{G_r : 0 < r < 1\}$ is equicontinuous on S. Hence by the Ascoli theorem, there exists a sequence $r_k \to 1$ such that G_{r_k} converges uniformly to a continuous function g on S.

Let

$$\epsilon_k = \sup_{\zeta \in S} |G(r_k\zeta) - \mathcal{P}[g](r_k\zeta)|.$$

Since $G(r_k\zeta) \to g(\zeta)$ uniformly, and by Proposition 5.6, $\mathcal{P}[g](r_k\zeta) \to g(\zeta)$ uniformly, $\epsilon_k \to 0$ as $k \to \infty$. Thus by the maximum principle,

$$G(z) \leq \mathcal{P}[g](z) + \epsilon_k, \qquad \text{for all } z, \ |z| \leq r_k.$$

If F is $\mathcal{M}$-harmonic, then

$$|G(z) - \mathcal{P}[g](z)| \le \epsilon_k, \qquad \text{for all } z, \ |z| \le r_k.$$

Letting $k \to \infty$, we have

$$(5.20) \qquad\qquad G(z) \le \mathcal{P}[g](z)$$

for all $z \in B$, with equality in (5.20) if F is $\mathcal{M}$-harmonic. Also, by (5.19) and Fatou's lemma,

$$\|g\|_p \le \|F\|_p, \qquad 1 \le p \le \infty.$$

To conclude the proof, let $\{h_j\}$ be a sequence of continuous functions on $\mathcal{U}$ which forms an approximate identity. For each j, let G_j be as defined by (5.17). Since F is continuous,

$$(5.21) \qquad\qquad \lim_{j \to \infty} G_j(z) = F(z), \qquad \text{for all } z \in B.$$

By the above, each $G_j = \mathcal{P}[g_j]$ for some $g_j \in L^p(S)$ with $\|g_j\|_p \le \|F\|_p$. If $p > 1$, then some subsequence of $\{g_j\}$ converges in the weak $*$ topology of $L^p(S)$ to some $f \in L^p(S)$. In particular, $\mathcal{P}[g_j] \to \mathcal{P}[f]$. Thus by (5.20) and (5.21), $F(z) \le \mathcal{P}[f](z)$, with equality if F is $\mathcal{M}$-harmonic. For the case $p = 1$, some subsequence of $\{g_j\}$ converges weak $*$ in the dual of $C(S)$. Thus $F(z) \le \mathcal{P}[\nu](z)$ for some finite signed measure ν on S. Again, if F is $\mathcal{M}$-harmonic, then equality holds for some complex measure ν on S. $\square$

5.4. The Dirichlet Problem for rB.

In Theorem 5.5 we proved that the Dirichlet problem for $\widetilde{\Delta}$ was solvable on B. The following theorem asserts that the Dirichlet problem for $\widetilde{\Delta}$ is also solvable for rB, where for $0 < r < 1$, $rB = \{rz : z \in B\}$. Similarly, let $rS = \{rt : t \in S\} = \{z : |z| = r\}$.

Theorem 5.9. ([Ru3]) *Fix r, $0 < r < 1$. If $f \in C(rS)$, then there exists $F \in C(r\overline{B})$ such that*
  (a) $\widetilde{\Delta}F = 0$ in rB,
  (b) $F(rt) = f(rt)$ for all $t \in S$, and
  (c) $F(0) = \int_S f(rt)\, d\sigma(t)$.

Proof. Let $\mathcal{H}(r\overline{B})$ denote the class of functions F wich are $\mathcal{M}$-harmonic on rB and continuous on $r\overline{B}$. For $F \in \mathcal{H}(r\overline{B})$, set

$$(5.22) \qquad \|F\| = \sup\{|F(w)| : w \in r\overline{B}\} = \sup\{|F(rt)| : t \in S\}.$$

The last identity follows from the maximum principle for $\mathcal{M}$-subharmonic functions. Let

$$\mathcal{H}(rS) = \{f \in C(rS) : f(rt) = F(rt) \text{ for some } F \in \mathcal{H}(r\overline{B})\}.$$

Then $\mathcal{H}(rS)$, with the sup norm, is a closed subspace of $C(rS)$ which by (5.22) is isomorphic to $\mathcal{H}(r\overline{B})$.

We now show that $\mathcal{H}(rS) = C(rS)$. If not, then by the Hahn-Banach theorem there exists a nontrivial linear functional γ on $C(rS)$ such that $\gamma(f) = 0$ for all $f \in \mathcal{H}(rS)$. Thus by the Riesz-representation theorem there exists a complex Borel measure ν on S such that

$$\int_S f(rt)\, d\nu(t) = 0 \qquad \text{for all } f \in \mathcal{H}(rS).$$

In particular,

$$\int_S \mathcal{P}[\psi](rt)\, d\nu(t) = 0 \qquad \text{for all } \psi \in C(S).$$

Thus by Fubini's theorem,

$$\int_S \mathcal{P}[\psi](rt)\, d\nu(t) = \int_S \mathcal{P}[\nu](r\zeta)\psi(\zeta)\, d\sigma(\zeta) = 0$$

for all $\psi \in C(S)$. Therefore $\mathcal{P}[\nu](r\zeta) = 0$ for all $\zeta \in S$. Hence by the maximum principle, $\mathcal{P}[\nu](z) = 0$ for all $z \in rB$. Since $\mathcal{M}$-harmonic functions are real analytic (see remark below), $\mathcal{P}[\nu](z) = 0$ for all $z \in B$. Thus by Corollary 5.7 $\nu = 0$.

Finally, to show that $F(0) = \int_S f(rt)\, d\sigma(t)$, define γ_0 on $\mathcal{H}(r\overline{B})$ by $\gamma_0(F) = F(0)$. This defines a bounded linear functional on $C(rS)$ in the obvious way. Thus there exists a measure ν on S such that

$$\int_S F(rt)\, d\nu(t) = F(0)$$

for all $F \in \mathcal{H}(r\overline{B})$. Let $\mu = \nu - \sigma$. Then if $\psi \in C(S)$,

$$\int_S \mathcal{P}[\mu](rt)\psi(t)\, d\sigma(t) = \int_S \left[\int_S \mathcal{P}(rt, \zeta)\, d\nu(\zeta) - \mathcal{P}(0, \zeta) \right] \psi(t)\, d\sigma(t)$$

$$= \int_S \left[\int_S \mathcal{P}(r\zeta, t)\, d\nu(\zeta) - \mathcal{P}(0, \zeta) \right] \psi(t)\, d\sigma(t) = 0.$$

Thus as above, $\mu = 0$, i.e., $\nu = \sigma$. $\square$

Remark. The fact that $\mathcal{M}$-harmonic functions are real analytic follows from the following theorem of Hörmander ([Ho, Theorem 7.5.1]): If L is an elliptic differential operator with real analytic coefficients, then every solution u of $Lu = 0$ is real analytic.

Proposition 5.10. *If F is a nonnegative $\mathcal{M}$-harmonic function on B, then there exists a nonnegative Borel measure ν on S such that $F(z) = \mathcal{P}[\nu](z)$.*

Proof. By Theorem 5.9

$$\int_S F(rt)\, d\sigma(t) = F(0) \qquad \text{for all } r, 0 < r < 1.$$

The result now follows by Theorem 5.8. $\square$

Proposition 5.11. *If f is $\mathcal{M}$-subharmonic on B, then for every $a \in B$,*

$$\int_S f(\varphi_a(rt))\, d\sigma(t)$$

is a nondecreasing function of r, $0 < r < 1$.

Proof. Without loss of generality we take $a = 0$. Suppose $0 < r_1 < r_2 < 1$. We suppose first that f is continuous. By Theorem 5.9, there exists a function G which is $\mathcal{M}$-harmonic on $r_2 B$ such that $G(r_2 t) = f(r_2 t)$. Thus by the maximum principle, $f(z) \leq G(z)$ for all $z \in r_2 B$. Therefore

$$\int_S f(r_1 t)\, d\sigma(t) \leq \int_S G(r_1 t)\, d\sigma(t) = G(0) = \int_S G(r_2 t)\, d\sigma(t) = \int_S f(r_2 t)\, d\sigma(t).$$

For arbitrary f, since f is uppersemicontinous on the compact set $r_2 S$, there exists a decreasing sequence $\{f_n\}$ of continuous functions $r_2 S$ such that $f_n(r_2 t) \to f(r_2 t)$. For each n, let $F_n \in \mathcal{H}(r_2 \overline{B})$ be such that $F_n(r_2 t) = f_n(r_2 t)$. Since $F_n(r_2 t) \geq f(r_2 t)$ for all $t \in S$, by the maximum principle $F_n(z) \geq f(z)$ for all $z \in r_2 B$. Therefore

$$\int_S f(r_1 t)\, d\sigma(t) \leq \int_S F_n(r_2 t)\, d\sigma(t) = \int_S f_n(r_2 t)\, d\sigma(t).$$

The result now follows by letting $n \to \infty$. $\square$

5.5. Remarks.

(1) If f is holomorphic (or pluriharmonic) on B, then for $0 < r < 1$, the function f_r is holomorphic (pluriharmonic) on $\overline{B}$. Thus by Corollary 5.2,

$$f_r(z) = \int_B \mathcal{P}(z,t) f_r(t)\, d\sigma(t), \qquad z \in B.$$

Using the above, it is now easy to show that if $f \in H^p(B)$ (or $\mathcal{P}h^p(B)$), $1 \leq p \leq \infty$, then some subsequence of $\{f_r\}$ converges weak-$\star$ to a function $f \in L^p(S)$, $p > 1$, or to a measure ν on S when $p = 1$, thereby giving the Poisson integral representation for such functions. These results are valid in any bounded symmetric domain D, where $\mathcal{P}$ is the Poisson-Szegö kernel of D, and S is the Bergman-Šilov boundary of B. For further details the reader is referred to [HM], [Sto1]. The existence of harmonic majorants for plurisubharmonic functions was considered by the author in [Sto3].

(2) For n-subharmonic (i.e. strongly subharmonic) functions on the polydisc U^n, the analogue of Theorem 5.8 was proved in [Ru1]. The Poisson-Szegö kernel for U^n is simply

$$(5.22) \qquad \mathcal{P}(z,t) = \prod_{k=1}^{n} \frac{(1 - |z_k|^2)}{|1 - z_k \bar{t}_k|^2}, \qquad t \in T^n, z \in U^n.$$

By the result of Furstenberg [Fu1], if f is a bounded weakly harmonic function on U^n, then f is strongly harmonic and there exists $f^* \in L^\infty(T^n)$ such that

$$f(z) = \int_{T^n} \mathcal{P}(z,t) f^*(t) \, d\sigma(t),$$

where here σ denotes normalized Lebesgue measure on T^n.

(3) For weakly harmonic functions, we also have an analogoue of Proposition 5.10, which for convenience we state in U^2. As in Section 3.2, for $\lambda = (\lambda_1, \lambda_2) \in \mathbb{R}^2$, $(z,t) \in U^2 \times T^2$, set

$$(5.23) \qquad \mathcal{P}^\lambda(z,t) = \mathcal{P}^{\lambda_1 + \frac{1}{2}}(z_1, t_1) \mathcal{P}^{\lambda_2 + \frac{1}{2}}(z_2, t_2).$$

If λ satisfies

$$\lambda_1^2 + \lambda_2^2 = \frac{1}{2} + c, \qquad c \geq -\frac{1}{2},$$

then for all $t \in T^n$,

$$\widetilde{\Delta} \mathcal{P}^\lambda(z,t) = 2c \, \mathcal{P}^\lambda(z,t).$$

Let S_c^+ denote the positive quadrant of the circle

$$\lambda_1^2 + \lambda_2^2 = \frac{1}{2} + c, \qquad c \geq -\frac{1}{2}.$$

The following theorem is a special case of a 1965 result of Karpelic [Ka]:

Theorem 5.12. *If F is a positive solution of $\widetilde{\Delta} F = 2c F$, $c \geq -\frac{1}{2}$, then there exists a positive measure ν on $T^2 \times S_c^+$ such that*

$$(5.24) \qquad F(z) = \int_{T^2 \times S_c^+} \mathcal{P}^\lambda(z,t) \, d\nu(t, \lambda).$$

The case $c = 0$ gives an integral representation for nonnegative weakly harmonic functions on U^2. Using the above result, Michelson [Mi] obtained a generalization of Furstenberg's result. For $\lambda \in S_c^+$ ($c \geq -\frac{1}{2}$), define

$$(5.25) \qquad \phi_\lambda(z) = \int_{T^2} \mathcal{P}^\lambda(z,t) \, d\sigma(t).$$

If $\widetilde{\Delta} F = 2c\, F$ and $|F(z)| \leq M\, \phi_\lambda(z)$ for all $z \in U^2$ and some positive constant M, then

$$F(z) = \int_{T^2} \mathcal{P}^\lambda(z,t) f(t)\, d\sigma(t)$$

for some f bounded by M. The case $\lambda = (\frac{1}{2}, \frac{1}{2})$ gives the result of Furstenberg.

(4) For weakly subharmonic functions in the polydisc, the analogue of Theorem 5.8 does not appear to be known.

6.
The Riesz Decomposition Theorem

One version of the classical Riesz decomposition theorem for subharmonic functions in the unit disc U is as follows: if f is subharmonic in U and f has a harmonic majorant in U, then

$$(6.1) \qquad f(z) = -\frac{1}{2\pi} \int_U \log \left| \frac{1 - \overline{w}z}{w - z} \right| \, d(\Delta f)(w) + H_f(z),$$

where H_f is the least harmonic majorant of f on U and Δf is the Riesz measure of f for the Laplacian Δ, i.e.,

$$\int_U \psi \, d(\Delta f) = \int_U f(z) \, \Delta \psi(z) \, dA(z), \qquad \psi \in C_c^2(U).$$

In U, $d\lambda(z) = \frac{1}{\pi}(1 - |z|^2)^{-2} \, dA(z)$, and

$$\widetilde{\Delta} = 2(1 - |z|^2)^2 \frac{\partial^2}{\partial z \partial \overline{z}} = \frac{1}{2}(1 - |z|^2)^2 \, \Delta.$$

Thus (6.1) can be rewritten as

$$(6.2) \qquad f(z) = - \int_U \log \left| \frac{1 - \overline{w}z}{w - z} \right| \, d\mu_f(w) + H_f(z),$$

where μ_f is the Riesz measure of f as defined by (4.15), namely

$$\int_B \psi \, d\mu_f = \int_B f \, \widetilde{\Delta} \psi \, d\lambda, \qquad \psi \in C_c^2(B).$$

The function

$$G(z, w) = \log \left| \frac{1 - \overline{w}z}{w - z} \right| = -\log |\varphi_w(z)|$$

is the **Green's** function of U, which satisfies

(a) $G(z, w) \geq 0$,

(b) for fixed $w \in U$, $z \to G(z, w)$ is harmonic on $U \sim \{w\}$, and superharmonic on U,

(c) for fixed $w \in U$, $z \to G(z,w) + \log|z - w|$ is harmonic on U, and

(d) for fixed $w \in U$, $\zeta \in T$,

$$\lim_{\substack{z \to \zeta \\ z \in U}} G(z,w) = 0.$$

In this chapter we will obtain the Green's function G for the operator $\tilde{\Delta}$ on B and prove the analogue of (6.2) for $\mathcal{M}$-subharmonic functions on B. As we will see, the function G will satisfy (a), (b), and (d), but not (c). The Riesz decomposition theorem for $\mathcal{M}$-subharmonic functions on B was proved by D. Ullrich in his dissertation[Ul1], and published in [Ul2]. Many of the results presented here are based on his work.

6.1. Harmonic Majorants for $\mathcal{M}$-Subharmonic Functions.

Definition. *An $\mathcal{M}$-subharmonic function f on B has an $\mathcal{M}$-**harmonic majorant** on B if there exists an $\mathcal{M}$-harmonic function h on B such that $f(z) \leq h(z)$ for all $z \in B$. Furthermore, if there exists an $\mathcal{M}$-harmonic function H satisfying*

(a) $f(z) \leq H(z)$, for all $z \in B$, and

(b) $H(z) \leq h(z)$ for any $\mathcal{M}$-harmonic majorant h of f,

*then H is called the least $\mathcal{M}$-**harmonic majorant** of f, and will be denoted by H_f.*

The following theorem gives necessary and sufficient conditions for the existence of the least $\mathcal{M}$-harmonic majorant.

Theorem 6.1. *Let $f \not\equiv -\infty$ be $\mathcal{M}$-subharmonic on B. Then the following are equivalent:*

(a) f has a least $\mathcal{M}$-harmonic majorant on B.

(b) f has an $\mathcal{M}$-harmonic majorant on B.

(c) $\lim_{r \to 1} \int_S f(rt)\, d\sigma(t) < \infty$.

For the proof of the theorem we will need the following two lemmas. The first of these is Harnack's inequality for $\mathcal{M}$-harmonic functions.

Lemma 6.2. *Let Ω be an open subset of B and $a \in \Omega$. If K is a compact subset of Ω, then there exists a constant C_K such that*

$$h(z) \leq C_K\, h(a)$$

for all $z \in K$ and all nonnegative harmonic functions h on Ω.

Proof. It suffices to consider the case $K = \overline{E(a,r)} \subset \Omega$. Choose r_1, $r < r_1 < 1$ such that

$$\overline{E(a,r)} \subset \overline{E(a,r_1)} \subset \Omega,$$

and choose $\delta > 0$ such that $E(z,\delta) \subset E(a,r_1)$ for all $z \in K$. Then by (4.3), if h is a nonnegative harmonic function on Ω,

$$h(z) = \frac{1}{\lambda(E(z,\delta))} \int_{E(z,\delta)} h(w)\,d\lambda(w)$$

$$\leq \frac{1}{\lambda(E(z,\delta))} \int_{E(z,r_1)} h(w)\,d\lambda(w) = \frac{\lambda(E(a,r_1))}{\lambda(E(z,\delta))} h(a)$$

for all $z \in K$. The result now follows with $C_K = \lambda(B(0,r_1))/\lambda(B(0,\delta))$. $\quad\square$

Lemma 6.3. *Let $f \not\equiv -\infty$ be $\mathcal{M}$-subharmonic on B. Then for each r, $0 < r < 1$, there exists an $\mathcal{M}$-harmonic function $F^{(r)}$ on rB such that*

(a) $f(z) \leq F^{(r)}(z)$ *for all z, $|z| < r$,*

(b) $\int_S f(rt)\,d\sigma(t) = F^{(r)}(0)$.

Furthermore, if F is harmonic on an open subset Ω of B with $r\overline{B} \subset \Omega$, and $F(z) \geq f(z)$ for all $z \in \Omega$, then

(c) $F^{(r)}(z) \leq F(z)$ *for all z, $|z| < r$.*

Proof. Fix r, $0 < r < 1$. As in Proposition 5.11 let $\{f_n\}$ be a decreasing sequence of continuous functions on rS which converges to $f(rt)$ for all $t \in S$. For each n, let $F_n \in \mathcal{H}(r\overline{B})$ be the $\mathcal{M}$-harmonic function such that $F_n(rt) = f_n(rt)$. The existence of such an F_n is assured by Theorem 5.9. By the maximum principle,

$$f(z) \leq F_{n+1}(z) \leq F_n(z) \qquad \text{for all } z, \ |z| < r.$$

For $|z| < r$, let
$$F^{(r)}(z) = \lim_{n \to \infty} F_n(z).$$

Clearly, $F^{(r)}(z) \geq f(z)$. Since the sequence $\{F_n\}$ is monotone decreasing, by Harnack's theorem (which as a consequence of Lemma 6.2 holds for $\mathcal{M}$-harmonic functions), the sequence $\{F_n\}$ either converges uniformly on compact subsets of rB, or $F^{(r)}(z) = -\infty$ for all $z \in rB$. Since f is locally integrable, $F^{(r)} \not\equiv -\infty$ on rB. Thus by uniform convergence, $F^{(r)}$ is $\mathcal{M}$-harmonic on rB. Finally, by the monotone convergence theorem,

(6.3) $F^{(r)}(0) = \lim_{n \to \infty} F_n(0) = \lim_{n \to \infty} \int_S f_n(rt)\,d\sigma(t) = \int_S f(rt)\,d\sigma(t).$

Suppose F is $\mathcal{M}$-harmonic on a domain $\Omega \supset r\overline{B}$ with $F(z) \geq f(z)$ for all $z \in rB$. With $\{f_n\}$ as above, for each n, let

$$g_n(rt) = \min\{f_n(rt), F(rt)\},$$

and let G_n be the corresponding $\mathcal{M}$-harmonic function in rB. Since $g_n(rt) \leq f_n(rt)$ for all $t \in S$, $G_n(z) \leq F_n(z)$ for all $z \in rB$. But $\{g_n\}$ is a decreasing sequence of continuous functions on rS with $\lim g_n(rt) = f(rt)$ for all $t \in S$. Thus by (6.3),

$$\lim_{n \to \infty} G_n(0) = \int_S f(rt)\, d\sigma(t) = \lim_{n \to \infty} F_n(0).$$

Therefore as a consequence of Lemma 6.2,

$$F^{(r)}(z) = \lim_{n \to \infty} F_n(z) = \lim_{n \to \infty} G_n(z).$$

However, by the maximum principle, $G_n(z) \leq F(z)$ for all n. Thus $F^{(r)}(z) \leq F(z)$ for all $z \in rB$. □

Proof of Theorem 6.1. Clearly (a) $\Rightarrow$ (b) $\Rightarrow$ (c). Suppose that (c) holds. Choose an increasing sequence $\{r_n\}$ with $r_n \to 1$. For each n, let $F^{(n)}$ be the $\mathcal{M}$-harmonic function in $r_n B$ satisfying the conclusions of Lemma 6.3. By part (c) of the lemma,

$$F^{(n)}(z) \leq F^{(n+1)}(z) \qquad \text{for all} \quad z \in r_n B.$$

Since

$$\lim_{n \to \infty} F^{(n)}(0) = \lim_{n \to \infty} \int_S f(r_n t)\, d\sigma(t) < \infty,$$

by Harnack's theorem

$$F(z) = \lim_{n \to \infty} F^{(n)}(z)$$

is $\mathcal{M}$-harmonic on B satisfying $F(z) \geq f(z)$ for all $z \in B$. Suppose h is a $\mathcal{M}$-harmonic majorant of f. Then by (c) of Lemma 6.2, $h(z) \geq F^{(n)}(z)$ for all $z \in r_n B$. Thus $F(z) \leq h(z)$ and therefore F is the least $\mathcal{M}$-harmonic majorant of f. □

Corollary 6.4. *Let $f \leq 0$ be $\mathcal{M}$-subharmonic on B with $f \not\equiv -\infty$. Then the least $\mathcal{M}$-harmonic majorant of f is the zero function if and only if*

$$\lim_{r \to 1} \int_S f(rt)\, d\sigma(t) = 0.$$

For a function f on B, let $f^+(z) = \max\{f(z), 0\}$. If f is $\mathcal{M}$-subharmonic on B, then so is f^+.

Theorem 6.5. *Let $f \not\equiv -\infty$ be $\mathcal{M}$-subharmonic on B. Then the following are equivalent:*

(a) $\sup_{0<r<1} \int_S f^+(rt)\, d\sigma(t) < \infty.$

(b) $f(z) \leq \mathcal{P}[\nu](z)$ *for some regular Borel measure ν on S.*

(c) f *has a least $\mathcal{M}$-harmonic majorant of the form $\mathcal{P}[\nu_f]$, for some regular Borel measure ν_f on S.*

Proof. Assume (a) holds. Since f^+ is $\mathcal{M}$-subharmonic on B, by Theorem 6.1 f^+ has a least $\mathcal{M}$-harmonic majorant H on B. Since H is nonnegative, by Proposition 5.10 $H(z) = \mathcal{P}[\nu](z)$ for some nonnegative Borel measure ν on S. Thus since

$$f(z) \leq f^+(z) \leq \mathcal{P}[\nu](z),$$

we obtain (b).

Assume that (b) holds, i.e., $f(z) \leq \mathcal{P}[\nu](z)$ for some regular Borel measure ν on S. Then for all r, $0 < r < 1$,

$$\int_S f(rt)\, d\sigma(t) \leq \int_S \mathcal{P}[\nu](rt)\, d\sigma(t) = \mathcal{P}[\nu](0)$$

which is finite. Hence $\lim_{r \to 1} \int_S f(rt)\, d\sigma(t) < \infty$. Thus by Theorem 6.1 f has a least $\mathcal{M}$-harmonic majorant H_f on B. Since H_f is the least $\mathcal{M}$-harmonic majorant of f,

$$H_f(z) \leq \mathcal{P}[\nu](z) \leq \mathcal{P}[\nu^+](z),$$

where ν^+ is the positive variation of ν. Therefore $H_f^+(z) \leq \mathcal{P}[\nu_+](z)$. Since $|H_f| = 2 H_f^+ - H_f$,

$$\int_S |H_f(rt)|\, d\sigma(t) = 2 \int_S H_f^+(rt)\, d\sigma(t) - \int_S H_f(rt)\, d\sigma(t) \leq 2\nu^+(S) - H_f(0).$$

Thus H_f satisfies (5.14), and as a consequence, by Theorem 5.8, $H_f = \mathcal{P}[\nu_f]$ for some regular Borel measure ν_f on S. The implication (c) $\Rightarrow$ (a) is obvious. $\square$

Definition. *The measure ν_f of Theorem 6.5 is called the **boundary measure** of the $\mathcal{M}$-subharmonic function f.*

6.2. The Green's Function for $\tilde{\Delta}$.

In this section we introduce the Green's function for the Laplace-Beltrami operator $\tilde{\Delta}$ on B. For $z \in B$, set

(6.4) $$g(z) = \frac{n+1}{2n} \int_{|z|}^1 (1 - t^2)^{n-1} t^{-2n+1}\, dt.$$

When $n = 1$, this becomes simply

$$g(z) = \log \frac{1}{|z|}.$$

Since the function g is radial, using the radial form (3.9) of $\widetilde{\Delta}$, one immediately obtains that $\widetilde{\Delta}g(z) = 0$ for all $z \in B, z \neq 0$. Thus g is $\mathcal{M}$-superharmonic on B and $\mathcal{M}$-harmonic on $B \sim \{0\}$. Since $(1-t^2)^{n-1}t^{-2n+1}$ is not integrable at 0 and integrable on $[r,1]$ for every $r > 0$,

$$g(0) = \infty \qquad \text{and} \qquad \lim_{|z| \to 1} g(z) = 0.$$

The following lemma gives some basic properties of g which will be needed later.

Lemma 6.6. *Let* $0 < \delta < \frac{1}{2}$ *be fixed. Then* g *satisfies the following:*

(6.5a) $\qquad g(z) \geq \dfrac{n+1}{4n^2}(1-|z|^2)^n \qquad$ *for all* $\quad z \in B$,

(6.5b) $\qquad g(z) \leq C_\delta (1-|z|^2)^n \qquad$ *for all* $\quad z \in B, |z| \geq \delta$,

where C_δ *is a positive constant depending only on* δ. *Furthermore, for all* $z, |z| \leq \delta$,

(6.6) $\qquad g(z) \approx \begin{cases} |z|^{-2n+2}, & n > 1, \\[2mm] \log \dfrac{1}{|z|}, & n = 1. \end{cases}$

In the above, $g(z) \approx h(z)$ means that there exist positive constants C_1, C_2 such that $C_1 h(z) \leq g(z) \leq C_2 h(z)$ for all indicated z. The proof is a routine estimation of the integral in (6.4), and thus is omitted.

Definition. *For* $z, w \in B$, *the (invariant)* **Green's function** G *for* $\widetilde{\Delta}$ *is defined by*

(6.7) $\qquad\qquad\qquad G(z,w) = g(\varphi_w(z))$

Since g is radial, $G(z,w) = G(w,z)$, and by the $\mathcal{M}$-invariance of $\widetilde{\Delta}$,

$$\widetilde{\Delta}_z G(z,w) = 0 \qquad \text{on} \quad B \sim \{w\}.$$

Although the above Green's function on B was first used extensively by Ullrich in his dissertation [Ul1] in 1981, the formula (6.4) had been derived much earlier in 1967 by K.T. Hahn and J. Mitchell [HM1].

We now proceed to show that G is indeed the fundamental solution for $\widetilde{\Delta}$.

Proposition 6.7. *For all $\psi \in C_c^2(B)$,*

$$(6.8) \qquad\qquad -\int_B g\, \tilde{\Delta}\psi \, d\lambda = \psi(0).$$

Proof. Since $-g$ is $\mathcal{M}$-subharmonic on B, by Theorem 4.10 there exists a nonnegative regular Borel measure μ_g on B such that

$$-\int_B g\, \tilde{\Delta}\psi \, d\lambda = \int_B \psi \, d\mu_g, \qquad \text{for all} \quad \psi \in C_c^2(B).$$

Since $\tilde{\Delta}g = 0$ on $B \sim \{0\}$, by Green's identity $\int_B \psi \, d\mu_g = 0$ for all $\psi \in C_c^2(B)$ such that $0 \notin \operatorname{supp}\psi$. Thus μ_g is a positive multiple of the point mass measure at 0. Therefore, there exists a positive constant c_n such that

$$-\int_B g\, \tilde{\Delta}\psi \, d\lambda = c_n\, \psi(0), \qquad \psi \in C_c^2(B).$$

The value of the constant c_n is not particularly important; this value could easily have been included in the definition of g as it was in [Ul2]. However, for completeness we now show that $c_n = 1$. To accomplish this we choose an appropriate ψ. Let

$$\psi(r) = (1 - r^2)^n.$$

Using the radial form (3.9) of $\tilde{\Delta}$ we obtain

$$\tilde{\Delta}\psi(r) = -\frac{4n^2}{n+1}(1 - r^2)^{n+1}.$$

Although our function ψ is C^2, it does not have compact support and thus technically formula (6.8) does not apply. At this point however we take the "eyes-closed" approach and just do it. The justification for doing so will follow from the Riesz decomposition theorem itself. Thus

$$\int_B g(z)\tilde{\Delta}\psi(z)\, d\lambda(z) = \int_B g(z)\tilde{\Delta}\psi(z)(1 - |z|^2)^{-n-1} \, d\nu(z)$$

$$= -\frac{4n^2}{n+1}\int_0^1 2n\, r^{2n-1} g(r)\, dr.$$

Since $g'(r) = -\frac{n+1}{2n}(1 - r^2)^{n-1}r^{-2n+1}$, an integration by parts yields

$$\int_B g(z)\tilde{\Delta}\psi(z)\, d\lambda(z) = \frac{4n^2}{n+1}\lim_{r\to 0+} r^{2n}g(r) - 2n\int_0^1 (1 - r^2)^{n-1}r\, dr$$

$$= \frac{4n^2}{n+1}\lim_{r\to 0+} r^{2n}g(r) - 1.$$

Finally, by L'Hopitals rule,

$$\lim_{r \to 0+} r^{2n} g(r) = \frac{n+1}{4n^2} \lim_{r \to 0} r^2 (1 - r^2)^{n-1} = 0.$$

Therefore,

$$-\int_B g(z) \tilde{\Delta} \psi(z) \, d\lambda(z) = 1 = c_n \psi(0) = c_n. \quad \square$$

Corollary 6.8. For all $\psi \in C_c^2(B)$,

(6.9) $\qquad -\int_B G(z,w) \tilde{\Delta} \psi(w) \, d\lambda(w) = -(g * \tilde{\Delta} \psi)(z) = \psi(z).$

Proof. Since $G(z,w) = g(\varphi_z(w))$, by the invariance of λ and $\tilde{\Delta}$,

$$\int_B G(z,w) \tilde{\Delta} \psi(w) \, d\lambda(w) = \int_B g(\varphi_z(w)) \tilde{\Delta} \psi(w) \, d\lambda(w)$$

$$= \int_B g(w)(\tilde{\Delta} \psi)(\varphi_z(w)) \, d\lambda(w)$$

$$= \int_B g(w) \tilde{\Delta}(\psi \circ \varphi_z)(w) \, d\lambda(w)$$

$$= -(\psi \circ \varphi_z)(0) = -\psi(z).$$

The last equality follows from (6.8) since $\psi \circ \varphi_z \in C_c^2(B)$. $\quad \square$

6.3. The Riesz Decomposition Theorem.

We are now ready to prove the Riesz decomposition theorem for $\mathcal{M}$-subharmonic functions on B. For the proof we will need the following lemma:

Lemma 6.9. *Suppose that* χ *is a* C^2 *radial function with compact support which satisfies* $\int_B \chi \, d\lambda = 0$. *Let* $v = -g * \chi$. *Then*

$$v \in C_c^2(B) \qquad \text{and} \qquad \tilde{\Delta} v = \chi.$$

Proof. Clearly v is C^2. We first show that v has compact support. For fixed z, the function $w \to G(z,w)$ is $\mathcal{M}$-harmonic in $B(0,|z|)$. Thus for all $\rho < |z|$,

$$\int_S G(z, \rho t) \, d\sigma(t) = G(z,0) = g(z).$$

Choose r, $0 < r < 1$, such that χ has support in $r\overline{B}$. Then for all z, $|z| > r$,

$$(g * \chi)(z) = \int_B G(z, w)\chi(w)\, d\lambda(w)$$

$$= 2n \int_0^r \rho^{2n-1}(1 - \rho^2)^{-n-1}\chi(\rho) \int_S G(z, \rho t)\, d\sigma(t)\, d\rho$$

$$= g(z) \int_B \chi(w)\, d\lambda(w) = 0.$$

Thus v has compact support.

Let $\psi \in C_c^2(B)$ be arbitrary. Then by Green's formula (3.3) and Fubini's theorem,

$$\int_B \psi(z)\widetilde{\Delta}v(z)\, d\lambda(z) = \int_B v(z)\widetilde{\Delta}\psi(z)\, d\lambda(z)$$

$$= -\int_B \int_B G(z, w)\chi(w)\, d\lambda(w)\, \widetilde{\Delta}\psi(z)\, d\lambda(z)$$

$$= -\int_B \chi(w) \int_B G(z, w)\widetilde{\Delta}\psi(z)\, d\lambda(z)\, d\lambda(w)$$

$$= \int_B \chi(w)\psi(w)\, d\lambda(w).$$

The last identity follows by Corollary 6.8. Therefore,

$$\int_B \psi(z)\widetilde{\Delta}v(z)\, d\lambda(z) = \int_B \psi(z)\chi(z)\, d\lambda(z).$$

Since v is C^2 and this holds for all $\psi \in C_c^2(B)$, $\widetilde{\Delta}v = \chi$. $\square$

Theorem 6.10. *Suppose $f \leq 0$, $f \not\equiv -\infty$, is $\mathcal{M}$-subharmonic in B satisfying*

$$(6.10) \qquad \lim_{r \to 1} \int_S f(rt)\, d\sigma(t) = 0.$$

Then

$$(6.11) \qquad f(z) = -\int_B G(z, w)\, d\mu_f(w),$$

where μ_f is the Riesz measure of f.

Proof. We first show that

$$f(0) = -\int_B g(w)\, d\mu_f(w).$$

Choose a sequence $\{r_j\}$ which decreases to 0. Let

$$A_j^1 = \{z : r_{j+1} \le |z| \le r_j\}, \qquad A_j^2 = \{z : 1 - r_j \le |z| \le 1 - r_{j+1}\}.$$

For $k = 1, 2$ and $j = 1, 2, ...$, let $\chi_j^k \ge 0$ be a C^2 radial function with support contained in A_j^k and satisfying $\int_B \chi_j^k d\lambda = 1$.

Since f is $\mathcal{M}$-subharmonic, by Proposition 4.7

$$f(0) = \lim_{j \to \infty} (f * \chi_j^1)(0) = \lim_{j \to \infty} \int_B f(w)\chi_j^1(w) \, d\lambda(w).$$

Also, for each j, if $\rho_j = 1 - r_j$,

$$\int_S f(\rho_j t) \, d\sigma(t) \le \int_B f(w)\chi_j^2(w) \, d\lambda(w) \le \int_S f(\rho_{j+1}t) \, d\sigma(t).$$

Thus by the hypothesis (6.10),

$$\lim_{j \to \infty} \int_B f(w)\chi_j^2(w) \, d\lambda(w) = 0.$$

Therefore, if $\chi_j = \chi_j^1 - \chi_j^2$, then

$$\lim_{j \to \infty} \int_B f(w)\chi_j(w) \, d\lambda(w) = f(0).$$

Furthermore, since g is $\mathcal{M}$-superharmonic on B, $g * \chi_j^1$ increases to g on B. Also, a similar argument as in Proposition 4.7 shows that $g * \chi_j^2$ decreases to zero. Thus $g * \chi_j$ increases to g. For each j, let $v_j = -g * \chi_j$. By Lemma 6.9, $v_j \in C_c^2(B)$ and $\widetilde{\Delta} v_j = \chi_j$. Thus by Theorem 4.10,

$$
\begin{aligned}
f(0) &= \lim_{j \to \infty} \int_B f(w)\chi_j(w) \, d\lambda(w) \\
&= \lim_{j \to \infty} \int_B f(w)\widetilde{\Delta} v_j(w) \, d\lambda(w) \\
&= \lim_{j \to \infty} \int_B v_j(w) \, d\mu_f(w) = -\lim_{j \to \infty} \int_B (g * \chi_j)(w) \, d\mu_f(w),
\end{aligned}
$$

which by the monotone convergence theorem

$$= -\int_B g(w) \, d\mu_f(w).$$

Let $z \in B$ be arbitrary and let $h = f \circ \varphi_z$. By Corollary 6.4, the least $\mathcal{M}$-harmonic majorant of f, and thus also of h, is the zero function. Thus h satisfies (6.10), and therefore

$$f(z) = h(0) = - \int_B g(w) \, d\mu_h(w),$$

where μ_h is the Riesz measure of h. But if $\psi \in C_c^2(B)$, by the invariance of $\widetilde{\Delta}$ and λ,

$$\int_B \psi \, d\mu_h = \int_B (f \circ \varphi_z) \widetilde{\Delta} \psi \, d\lambda$$

$$= \int_B f \widetilde{\Delta}(\psi \circ \varphi_z) \, d\lambda = \int_B (\psi \circ \varphi_z) \, d\mu_f.$$

As a consequence,

$$\int_B \psi \, d\mu_h = \int_B (\psi \circ \varphi_z) \, d\mu_f.$$

for any nonnegative Borel measurable function ψ. Therefore,

$$f(z) = - \int_B g(\varphi_z(w)) \, d\mu_f(w) = - \int_B G(z, w) \, d\mu_f(w). \quad \square$$

Corollary 6.11. *(Riesz Decomposition Theorem) Suppose $f \not\equiv -\infty$ is $\mathcal{M}$-subharmonic on B and has an $\mathcal{M}$-harmonic majorant on B. Then*

(6.12) $$f(z) = H_f(z) - \int_B G(z, w) \, d\mu_f(w),$$

where μ_f is the Riesz measure of f and H_f is the least $\mathcal{M}$-harmonic majorant of f.

Proof. Let $H_f(z)$ be least $\mathcal{M}$-harmonic majorant of f, and let $h(z) = f(z) - H_f(z)$. Then h satisfies (6.10). Furthermore, since H_f is $\mathcal{M}$-harmonic, $\mu_h = \mu_f$, which proves the result. $\square$

Remark: We can now justify our computation of the constant c_n in Proposition 6.7. Since $\psi(r) = (1 - r^2)^n$ satisfies $\widetilde{\Delta}\psi \le 0$, the function ψ is $\mathcal{M}$-superharmonic on B and satisfies (6.10). Thus

$$\psi(0) = \int_B g(w) \, d\mu_\psi(w).$$

But since ψ is C^2, the Riesz measure of ψ is given by $d\mu_\psi = -\widetilde{\Delta}\psi \, d\lambda$. Thus

$$\psi(0) = - \int_B g \widetilde{\Delta}\psi \, d\lambda.$$

6.4. Green Potentials.

Definition. *If μ is a nonnegative regular Borel measure on B, the function G_μ defined by*

$$(6.13) \qquad G_\mu(z) = \int_B G(z,w)\,d\mu(w)$$

is called the (invariant) **Green potential** *of μ, provided $G_\mu \not\equiv +\infty$.*

If $G_\mu \not\equiv \infty$, then since $z \to G(z,w)$ is $\mathcal{M}$-superharmonic on B for each $w \in B$, by Tonelli's theorem so is the function $G_\mu(z)$. Furthermore, G_μ is $\mathcal{M}$-harmonic on $B \sim S_\mu$, where S_μ is the support of the measure μ. We will shortly see that G_μ is not identically $+\infty$ if and only if the measure μ satisifies

$$(6.14) \qquad \int_B (1 - |w|^2)^n \, d\mu(w) < \infty.$$

Theorem 6.12. *Let $f \not\equiv -\infty$ be $\mathcal{M}$-subharmonic on B and let μ_f be the Riesz measure of f. Then f has an $\mathcal{M}$-harmonic majorant on B if and only if there exists $t_o \in B$ such that*

$$(6.15) \qquad \int_B G(t_o, z)\,d\mu_f(z) < \infty.$$

Proof. Suppose f has an $\mathcal{M}$-harmonic majorant. Then by the Riesz decomposition theorem

$$f(z) = H_f(z) - \int_B G(z,w)\,d\mu_f(w),$$

where H_f is the least $\mathcal{M}$-harmonic majorant of f on B. Since $f(z) > -\infty$ a.e. on B, we obtain (6.15) for a.e. $t_o \in B$.

Conversely, suppose the Riesz measure μ_f of f satisfies (6.15) for some $t_o \in B$. Set

$$(6.16) \qquad V_f(z) = \int_B G(z,w)\,d\mu_f(w).$$

Since $V_f(t_o) < \infty$, V_f is $\mathcal{M}$-superharmonic on B. Set $h(z) = f(z) + V_f(z)$, which by the remark following (4.3) is defined a.e. on B and is locally integrable. Let $\psi \in C_c^2(B)$, $\psi \geq 0$, and consider $\int h \widetilde{\Delta} \psi \, d\lambda$. By Corollary 6.8,

$$\int_B G(z,w)\widetilde{\Delta}\psi(w)\,d\lambda(w) = (g * \widetilde{\Delta}\psi)(z) = -\psi(z).$$

Thus by Fubini's Theorem and the definition of μ_f,

$$\int_B V_f \widetilde{\Delta}\psi \, d\lambda = -\int_B \psi \, d\mu_f = -\int_B f \widetilde{\Delta}\psi \, d\lambda.$$

Therefore $\int_B h\widetilde{\Delta}\psi \, d\lambda = 0$ for all $\psi \in C_c^2(B)$ with $\psi \geq 0$. Thus by Corollary 4.9, there exists an $\mathcal{M}$-harmonic function H on B such that $f + V_f = H$ a.e. on B. Since $V_f \geq 0$, we have that $f(w) \leq H(w)$ a.e.. However, as a consequence of (4.3), if $z \in B$, and $r > 0$ is sufficiently small,

$$f(z) \leq \frac{1}{\lambda(E(z,r))} \int_B f(w) \, d\lambda(w) \leq \frac{1}{\lambda(E(z,r))} \int_B H(w) \, d\lambda(w) = H(z).$$

Therefore, the inequality holds for all $z \in B$, and thus f has an $\mathcal{M}$-harmonic majorant on B. $\square$

For $t \in B$ and $\epsilon > 0$, let

(6.17) $\Omega_\epsilon(t) = \{z \in B : G(t,z) \geq \epsilon\}.$

Proposition 6.13. *Let $f \not\equiv -\infty$ be $\mathcal{M}$-subharmonic in B with an $\mathcal{M}$-harmonic majorant, and let μ_f be the Riesz measure f. Then*

(6.18) $\displaystyle\lim_{s \to 0^+} s \, \mu_f(\Omega_s(t)) = 0 \quad$ *for all* $\quad t \in B.$

Proof. Let $t_o \in B$ be arbitrary, and let V_f be the $\mathcal{M}$-superharmonic function on B defined by (6.16). Since $s \to \mu_f(\Omega_s(t_o))$ is the distribution function of $G(t_o, w)$,

$$\int_0^\infty \mu_f(\Omega_s(t_o)) \, ds = \int_B G(t_o, w) \, d\mu_f(w) = V_f(t_o).$$

Therefore,

$$\epsilon \mu_f(\Omega_\epsilon(t_o)) \leq \int_0^\epsilon \mu_f(\Omega_s(t_o)) \, ds \to 0 \quad \text{as } \epsilon \to 0^+$$

whenever $V_f(t_o) < \infty$.

Suppose $V_f(t_o) = \infty$. Fix $\alpha > 0$ and let ν_α be the measure defined by $(1 - \chi_\alpha)\mu_f$, where χ_α is the characteristic function of $\Omega_\alpha(t_o)$. Let V_α denote the corresponding function defined by (6.13) for the measure ν_α. Since V_α is $\mathcal{M}$-harmonic at t_o, $V_\alpha(t_o) < \infty$, and hence as above, $\epsilon\nu_\alpha(\Omega_\epsilon(t_o)) \to 0$ as $\epsilon \to 0$. Since $\Omega_\alpha(t_o)$ is compact, $\mu_f(\Omega_\alpha(t_o))$ is finite. Thus since

$$\epsilon \mu_f(\Omega_\epsilon(t_o)) = \epsilon \mu_f(\Omega_\alpha(t_o)) + \epsilon\nu_\alpha(\Omega_\epsilon(t_o)),$$

$\lim_{\epsilon \to 0} \epsilon \mu_f(\Omega_\epsilon(t_o)) = 0$, which proves the result. $\square$

Theorem 6.14. Let $f \not\equiv -\infty$ be $\mathcal{M}$-subharmonic on B with Riesz measure μ_f. Then f has an $\mathcal{M}$-harmonic majorant on B if and only if

$$(6.19) \qquad \int_B (1 - |z|^2)^n \, d\mu_f(z) < \infty.$$

Furthermore, if this is the case, then

$$(6.20) \qquad \lim_{r \to 1} (1 - r^2)^n \, \mu_f(B(0, r)) = 0.$$

Proof. Suppose f has an $\mathcal{M}$-harmonic majorant, and $t_o \in B$ is such that (6.15) holds. As a consequence of (6.5a)

$$G(t_o, w) \geq C_n (1 - |\varphi_{t_o}(w)|^2)^n.$$

Thus by identity (1.16),

$$(6.21) \quad G(t_o, w) \geq C_n \frac{(1 - |t_o|^2)^n (1 - |w|^2)^n}{|1 - \langle t_o, w \rangle|^{2n}} \geq C_n \left(\frac{1 - |t_o|}{1 + |t_o|} \right)^n (1 - |w|^2)^n$$

for all $w \in B$. Thus μ_f satisfies (6.19).

Conversely, suppose (6.19) holds. Let $0 < c < 1$ be fixed, and let V_1 and V_2 be defined by

$$V_1(z) = \int_{cB} G(z, w) \, d\mu_f(w),$$

$$V_2(z) = \int_{B \sim cB} G(z, w) \, d\mu_f(w).$$

where $cB = B(0, c)$. Since $\mu_{f|cB}$ is a finite measure, V_1 is $\mathcal{M}$-superharmonic on B. Also, by (6.5b), there exists a constant C such that

$$G(0, w) \leq C (1 - |w|^2)^n \qquad \text{for all} \quad |w| \geq c.$$

Thus $V_2(0) < \infty$ by hypothesis, and therefore V_2 is also $\mathcal{M}$-superharmonic on B. Thus $V_f = V_1 + V_2$ is $\mathcal{M}$-superharmonic on B and consequently finite a.e. Therefore μ_f satisfies (6.15) for a.e. $t_o \in B$.

Finally, to obtain (6.20) it suffices to take $t = 0$ in (6.18). Since $G(0, z) \geq C_n (1 - |z|^2)^n$ for all $z \in B$, $B_r \subset \Omega_s(0)$ where $s = C_n (1 - r^2)^n$. Thus (6.20) now follows from (6.18). $\square$

Remark: The above proof shows that if μ is a nonnegative regular Borel measure on B, then the potential G_μ is $\mathcal{M}$-superharmonic on B if and only if μ satisfies (6.14).

Definition. *A nonnegative $\mathcal{M}$-superharmonic function $V \not\equiv \infty$ is called an* **invariant potential** *on B if there exists a regular Borel measure μ on B such that $V = G_\mu$.*

For the proof of Proposition 6.16 and later results, we require the following lemma:

Lemma 6.15. *If K is a compact subset of B, then there exists a positive constant C_K and an r_o, $0 < r_o < 1$, such that*

$$(6.22) \qquad G(z,w) \leq C_K (1 - |z|^2)^n (1 - |w|^2)^n,$$

for all $w \in K$, and all $z \in B$ with $|z| \geq r_o$.

Proof. Fix δ, $0 < \delta < \frac{1}{2}$. For this δ, since K is compact, there exists r_o, $0 < r_o < 1$, such that

$$\bigcup_{w \in K} E(w, \delta) \subset B(0, r_o).$$

Thus if $w \in K$ and $|z| \geq r_o$, $|\varphi_z(w)| \geq \delta$. Therefore by inequality (6.5b) and identity (1.16),

$$G(z, w) \leq C_\delta (1 - |\varphi_z(w)|^2)^n = C_\delta \frac{(1 - |z|^2)^n (1 - |w|^2)^n}{|1 - \langle z, w\rangle|^{2n}}.$$

But since K is compact, there exists $c_K > 0$ such that

$$|1 - \langle z, w\rangle| \geq (1 - |z||w|) \geq (1 - |w|) \geq c_K$$

for all $w \in K$. The result now follows with $C_K = C_\delta / c_K^n$. $\square$

Proposition 6.16. *A nonnegative $\mathcal{M}$-superharmonic function $V \not\equiv \infty$ is an* **invariant potential** *on B if and only if*

$$(6.23) \qquad \lim_{r \to 1} \int_S V(rt)\, d\sigma(t) = 0.$$

Proof. If V satisfies (6.23), then by the Riesz decomposition theorem V is the Green potential of it's Riesz measure.

Conversely, suppose $V = G_\mu$ for some nonnegative regular Borel measure μ satisfying (6.14). Fix δ, $0 < \delta < \frac{1}{2}$, and suppose $0 < \rho < 1$. Write

$$V(z) = \int_{\rho B} G(z, w)\, d\mu(w) + \int_{B \sim \rho B} G(z, w)\, d\mu(w).$$

Since $z \to G(z, w)$ is $\mathcal{M}$-superharmonic for fixed $w \in B$, $|w| \geq \delta$,

$$\int_S \int_{B \sim \rho B} G(rt, w) \, d\mu(w) \, d\sigma(t) = \int_{B \sim \rho B} \int_S G(rt, w) \, d\sigma(t) \, d\mu(w)$$

$$\leq \int_{B \sim \rho B} G(0, w) \, d\mu(w)$$

$$\leq C_\delta \int_{B \sim \rho B} (1 - |w|^2)^n \, d\mu(w).$$

The last inequality follows from (6.5b). Since $\int_B (1 - |w|^2)^n d\mu(w) < \infty$,

$$\lim_{\rho \to 1} \int_{B \sim \rho B} (1 - |w|^2)^n \, d\mu(w) = 0.$$

Thus, given $\epsilon > 0$, there exists ρ such that

$$C_\delta \int_{B \sim \rho B} (1 - |w|^2)^n \, d\mu(w) < \epsilon.$$

For this ρ, by the preceeding lemma, there exists r_o and a constant $C\rho$ such that

$$G(z, w) \leq C_\rho (1 - |z|^2)^n (1 - |w|^2)^n$$

for all $w \in \rho B$, $|z| \geq r_o$. Therefore, for all $r \geq r_o$,

$$\int_S V(rt) \, d\sigma(t) \leq C_\rho (1 - r^2)^n \int_{\rho B} (1 - |w|^2)^n \, d\mu(w) + \epsilon.$$

Thus

$$\limsup_{r \to 1} \int_S V(rt) \, d\sigma(t) \leq \epsilon,$$

from which the result now follows. $\square$

6.5. A Characterization of H^p Spaces.

We now use Theorem 6.14 to give a characterization of the Hardy H^p spaces. Recall that for $0 < p < \infty$, $H^p(B)$ denotes the set of holomorphic functions f on B satisfying

$$(6.24) \qquad \sup_{0 < r < 1} \int_S |f(rt)|^p \, d\sigma(t) < \infty.$$

It was observed in [Zy, p. 208] that a holomorphic function $f \in H^2(U)$ if and only if

$$\int_U (1 - |z|) |f'(z)|^2 \, dx dy < \infty.$$

This result was extended to all p, $0 < p < \infty$, by S. Yamashita [Ya] who proved that $f \in H^p(U)$ if and only if

$$(6.25) \qquad \int_U (1 - |z|) |f(z)|^{p-2} |f'(z)|^2 \, dx dy < \infty.$$

This result extends to $H^p(B)$ as follows:

Theorem 6.17. *([Sto 8]) A holomorphic function f is an element of $H^p(B)$, $0 < p < \infty$, if and only if*

$$(6.26) \qquad \int_B (1 - |z|^2)^n |f(z)|^{p-2} |\widetilde{\nabla} f(z)|^2 \, d\lambda(z) < \infty,$$

where $\widetilde{\nabla}$ denotes the invariant gradient defined by (3.19). Futhermore, if $f \in H^p(B)$ then

$$(6.27) \qquad \lim_{r \to 1} (1 - r^2)^n \int_{B_r} |f(z)|^{p-2} |\widetilde{\nabla} f(z)|^2 \, d\lambda(z) = 0,$$

where $B_r = \{z \in B : |z| < r\}$.

Remark: When $n = 1$, $|\widetilde{\nabla} f(z)|^2 = (1 - |z|^2)^2 |f'(z)|^2$, and (6.26) reduces to (6.25).

Proof. Since $|f(z)|^p$ $(0 < p < \infty)$ is plurisubharmonic on B, it is also $\mathcal{M}$-subharmonic. Thus by Theorem 6.5, $f \in H^p(B)$ if and only if $|f|^p$ has an $\mathcal{M}$-harmonic majorant on B. By Theorem 6.14 this is the case if and only if the Riesz measure of $|f|^p$ satisfies (6.19). Thus to prove the result, we will show that the Riesz measure of $|f(z)|^p$, $0 < p < \infty$, is given by $f_p^\sharp \, d\lambda$, where

$$(6.28) \qquad f_p^\sharp(z) = \tfrac{1}{2} p^2 |f(z)|^{p-2} |\widetilde{\nabla} f(z)|^2.$$

Let $Z_f = \{z \in B : f(z) = 0\}$. Since f is holomorphic with $f \not\equiv 0$, Z_f has measure zero and thus $f_p^\sharp$ is defined a.e. on B. On $B \sim Z_f$, $\widetilde{\Delta} |f(z)|^p = f_p^\sharp(z)$.

For $\epsilon > 0$, let $g_\epsilon(z)$ be defined by

$$g_\epsilon(z) = (|f(z)|^2 + \epsilon)^{\frac{p}{2}}.$$

Then g_ϵ is C^∞ and $g_\epsilon(z) \to |f(z)|^p$ uniformly on compact subsets of B. A straightforward computation gives

$$\frac{\partial^2 g_\epsilon}{\partial z_i \partial \bar{z}_j} = (\tfrac{p}{2})^2 (|f(z)|^2 + \epsilon)^{\frac{p}{2}-1} \left[\frac{|f(z)|^2 + \frac{2}{p}\epsilon}{|f(z)|^2 + \epsilon} \right] \frac{\partial f}{\partial z_i} \frac{\partial \bar{f}}{\partial \bar{z}_j}.$$

Thus by (3.7) and (3.19),

$$\widetilde{\Delta} g_\epsilon(z) = \tfrac{1}{2} p^2 (|f(z)|^2 + \epsilon)^{\frac{p}{2}-1} \left[\frac{|f(z)|^2 + \frac{2}{p}\epsilon}{|f(z)|^2 + \epsilon} \right] |\widetilde{\nabla} f(z)|^2.$$

For $p \geq 2$, $\widetilde{\Delta} g_\epsilon(z) \to f_p^\sharp(z)$ as $\epsilon \to 0$ for all $z \in B$, and for $0 < p < 2$, the convergence is a.e.. Also, from the inequality

$$\frac{t + \frac{2}{p}\epsilon}{t + \epsilon} \leq \max\{\tfrac{2}{p}, 1\} = C_p$$

valid for all $t \in [0, \infty)$, we obtain

$$(6.29) \qquad \widetilde{\Delta} g_\epsilon(z) \leq \tfrac{1}{2} p^2 C_p \left(|f(z)|^2 + \epsilon \right)^{\frac{p}{2}-1} |\widetilde{\nabla} f(z)|^2.$$

We now show that $f_p^\sharp$ is locally integrable on B. Let K be any compact subset of B, and let $\psi \in C_c^2(B)$ with $\psi \geq 0$ be such that $\psi(z) = 1$ for all $z \in K$. By Fatou's lemma,

$$\int_K f_p^\sharp(z) \, d\lambda(z) \leq \liminf_{\epsilon \to 0^+} \int_K \widetilde{\Delta} g_\epsilon(z) \, d\lambda(z).$$

But

$$\int_K \widetilde{\Delta} g_\epsilon \, d\lambda \leq \int_B \psi \, \widetilde{\Delta} g_\epsilon \, d\lambda = \int_B g_\epsilon \, \widetilde{\Delta} \psi \, d\lambda.$$

In the above, the last equality follows by Green's identity. Since $\widetilde{\Delta}\psi$ is continuous with compact support and $g_\epsilon \to |f|^p$ uniformly on compact subsets of B, we have that $f_p^\sharp$ is locally integrable on B. Finally, to prove the result it remains to be shown that

$$(6.30) \qquad \int_B |f|^p \widetilde{\Delta}\psi \, d\lambda = \int_B \psi f_p^\sharp \, d\lambda$$

for all $\psi \in C_c^2(B)$. Fix such a ψ. If $p \geq 2$, $\widetilde{\Delta} g_\epsilon$ is uniformly bounded on compact sets, and if $0 < p < 2$, by (6.29)

$$\widetilde{\Delta} g_\epsilon(z) \leq \tfrac{2}{p} f_p^\sharp(z) \qquad \text{a.e. on } B.$$

Thus since $\widetilde{\Delta} g_\epsilon \to f_p^\sharp$ a.e. on B, and $f_p^\sharp$ is locally integrable on B, by Lebesgue's dominated convergence theorem and Green's identity,

$$\int_B \psi f_p^\sharp \, d\lambda = \lim_{\epsilon \to 0^+} \int_B \psi \, \widetilde{\Delta} g_\epsilon \, d\lambda$$
$$= \lim_{\epsilon \to 0^+} \int_B g_\epsilon \, \widetilde{\Delta} \psi \, d\lambda = \int_B |f|^p \widetilde{\Delta}\psi \, d\lambda,$$

which proves (6.30). Therefore, the Riesz measure of $|f|^p$ is given by (6.28), which establishes (6.26). The result (6.27) now follows from (6.20). $\square$

For the space $\mathcal{H}^p(B)$ of $\mathcal{M}$-harmonic functions u on B satisfying

$$\sup_{0 < r < 1} \int_B |u(rt)|^p \, d\sigma(t) < \infty.$$

we obtain the following analogous result:

Theorem 6.18. *Let $1 < p < \infty$. Then $u \in \mathcal{H}^p(B)$ if and only if*

$$
(6.31) \qquad \int_B (1 - |z|^2)^n |u(z)|^{p-2} |\widetilde{\nabla} u(z)|^2 \, d\lambda(z) < \infty,
$$

where $\widetilde{\nabla} u$ is defined by (3.16). Furthermore, if $u \in \mathcal{H}^p(B)$, then

$$
(6.32) \qquad \lim_{r \to 1} (1 - r^2)^n \int_{B_r} |u(z)|^{p-2} |\widetilde{\nabla} u(z)|^2 \, d\lambda(z) = 0.
$$

Proof. As in the proof of Theorem 6.17, we will show that for $1 < p < \infty$, the invariant Riesz measure of $|u|^p$ is given by $u_p^* d\lambda$, where

$$
(6.33) \qquad u_p^*(z) = p(p-1)|u(z)|^{p-2} |\widetilde{\nabla} u(z)|^2.
$$

As in the previous theorem, for $\epsilon > 0$, set

$$
u_\epsilon(z) = (u(z)^2 + \epsilon)^{p/2}.
$$

By computation,

$$
\frac{\partial^2 u_\epsilon}{\partial z_j \partial \bar{z}_j} = p(u^2 + \epsilon)^{\frac{p}{2}-1} \left[\frac{(p-1)u^2 + \epsilon}{u^2 + \epsilon} \right] \frac{\partial u}{\partial z_i} \frac{\partial u}{\partial \bar{z}_j} + p(u^2 + \epsilon)^{\frac{p}{2}-1} u \frac{\partial^2 u}{\partial z_i \partial \bar{z}_j}.
$$

Since u is real valued, $\frac{\partial u}{\partial \bar{z}_j} = \overline{\frac{\partial u}{\partial z_j}}$, and since u is $\mathcal{M}$-harmonic, $\widetilde{\Delta} u = 0$. Thus by (3.7) and (3.16)

$$
\widetilde{\Delta} u_\epsilon(z) = p(u(z)^2 + \epsilon)^{\frac{p}{2}-1} \left[\frac{(p-1)u(z)^2 + \epsilon}{u(z)^2 + \epsilon} \right] |\widetilde{\nabla} u(z)|^2.
$$

Let $Z_u = \{ z \in B : u(z) = 0 \}$. Since $\mathcal{M}$-harmonic functions are real analytic, Z_u also has Lebesgue measure 0. Therefore

$$
\lim_{\epsilon \to 0} \widetilde{\Delta} u_\epsilon(z) = u_p^*(z)
$$

a.e. on B. An argument identical to that of the preceeding theorem shows that $u_p^* \in L^1_{loc}$, and that

$$
\int_B |u|^p \, \widetilde{\Delta} \psi \, d\lambda = \int_B u_p^* \psi \, d\lambda
$$

for all $\psi \in C_c^2(B)$. Thus the Riesz measure of $|u|^p$ is given by $u_p^* d\lambda$. $\square$

6.6. Remarks.

We conclude this chapter with some remarks concerning plurisubharmonic functions. For a domain $\Omega \subset \mathbb{C}^n$, the pluri-complex Green function $U_\Omega(z,w)$ with pole at $w \in \Omega$ is defined by

$$(6.34) \qquad U_\Omega(z,w) = \sup\{u(z)\}$$

where the supremum is taken over all p.s.h. functions $u \leq 0$ on Ω such that $u(z) - \log|z - w|$ is bounded above in a neighborhood of w. In the case of the unit ball B, U_B is given by

$$(6.35) \qquad U_B(z,w) = \log|\varphi_z(w)|.$$

Using techniques similar to those used in Proposition 6.16, one can prove that if μ is a nonnegative regular Borel measure on B satisfying

$$(6.36) \qquad \int_B (1 - |w|^2)\, d\mu(w) < \infty,$$

then the function V_μ defined by

$$(6.37) \qquad V_\mu(z) = \int_B \log|\varphi_z(w)|\, d\mu(w)$$

is a plurisubharmonic function on B whose least pluriharmonic majorant is the zero function. The converse however is false. S-Y Li has given an example of a p.s.h. function on B whose least pluriharmonic majorant is zero, but which cannot be expressed in the form (6.37) when $n > 1$. This however still leaves open the question of what are sufficient conditions on a plurisubharmonic function $V \leq 0$ such that V is the pluri-complex Green potential of a measure μ.

There is an analogue of the Riesz decomposition theorem for continuous p.s.h. functions due to J-P Demailly for hyperconvex domains in $\mathbb{C}^n$. A bounded domain Ω is **hyperconvex** if there exists a function $\varphi : \Omega \to [-1,0)$ which is a continuous p.s.h. exhaustion function of Ω. To state the result of Demailly, we need to introduce the Monge-Ampere operator in $\mathbb{C}^n$. For $j = 1, ..., n$, let

$$dz_j = dx_j + dy_j, \qquad d\bar{z}_j = dx_j - idy_j.$$

Then the volume form dV on $\mathbb{C}^n$ becomes

$$dV(z) = \left(\frac{i}{2}\right)^n \prod_{j=1}^n dz_j \wedge d\bar{z}_j$$

$$= dx_1 \wedge dy_1 \wedge \cdots \wedge dx_n \wedge dy_n.$$

The ∂ and $\bar{\partial}$ operators are defined by

$$\partial = \sum_{j=1}^{n} \frac{\partial}{\partial z_j} dz_j, \qquad \bar{\partial} = \sum_{j=1}^{n} \frac{\partial}{\partial \bar{z}_j} d\bar{z}_j.$$

Let $d = \partial + \bar{\partial}$ and $d^c = i(\bar{\partial} - \partial)$. The operator $(dd^c)^n$ defined by

$$(dd^c u)^n = dd^c u \wedge \cdots \wedge dd^c u$$

$$= 4^n n! \det \left(\frac{\partial^2 u}{\partial z_i \partial \bar{z}_j} \right) dV$$

for $u \in C^2$ is called the **Monge-Ampere** operator in $\mathbb{C}^n$. If u is pluri-harmonic, then we clearly have $(dd^c u)^n = 0$; however, these are not the only functions annihilated by $(dd^c)^n$. The following two theorems, which are stated without proof, are due to Demailly [De].

Theorem 6.19. *If Ω is a bounded hyperconvex domain in $\mathbb{C}^n$ with pluri-complex Green function U_Ω, then*

(a) $U_\Omega(z, w) \to 0$ *as* $z \to \partial\Omega$, *and*

(b) $(dd^c U_\Omega(\cdot, w))^n = (2\pi)^n \delta_w$,

where δ_w is pointmass at w.

Theorem 6.20. *Suppose V is p.s.h. on Ω, continuous on $\bar{\Omega}$, where Ω is a bounded hyperconvex domain in $\mathbb{C}^n$, $n \geq 2$. For each $w \in \Omega$, there exists a measure ν_w on $\partial\Omega$ such that*

$$(6.38) \quad V(w) = \int_{\partial\Omega} V \, d\nu_w - \frac{1}{(2\pi)^n} \int_\Omega (dd^c V) \wedge (dd^c \varphi)^{n-2} \wedge d\varphi \wedge d^c \varphi$$

where $\varphi = U_\Omega(\cdot, w)$.

In the case of the unit ball B, the measure ν_w is given by

$$d\nu_w(t) = \mathcal{P}(w, t) \, d\sigma(t).$$

For further information on the pluri-complex Green function and the Monge-Ampere operator the reader is referred to the following references: [BT], [Ce], [De], [Kl1, Kl2], among many others.

7.
Admissible Boundary
Limits of Poisson Integrals

The classical theorem of Fatou [Fa] states that if f is a harmonic function in the unit disc U which is the Poisson integral of a measure ν on T, then f has nontangential limits at almost every $t \in T$. More precisely, for every $\alpha > 1$,

$$\lim_{\substack{z \to t \\ z \in \Gamma_\alpha(t)}} f(z) \quad \text{exists for a.e.} \quad t \in T,$$

where for $\alpha > 1$ and $t \in T$, $\Gamma_\alpha(t)$ is the nontangential approach region with vertex t defined by

$$\Gamma_\alpha(t) = \{z \in U : |z - t| < \frac{\alpha}{2}(1 - |z|^2)\}.$$

In this chapter we will prove the analogou of Fatou's theorem for $\mathcal{M}$-harmonic functions on the ball.

The first extension of Fatou's theorem to several complex variables dates back to 1939 to the work of J. Marcinkiewicz and A. Zygmund dealing with Poisson integrals on the polydisc U^n ([Zy, Chapter XVII]). If F is the Poisson integral of $f \in L^p(T^n)$, $p > 1$, then

$$\lim F(z) = f(t) \quad \text{a.e.} \quad t \in T^n$$

provided that for each j, $z_j \to t_j$ nontangentially. However, for $p = 1$, this fails in general and remains only true under the additional assumption that there exists a positive constant C such that

$$\frac{1}{C} < \frac{1 - |z_j|}{1 - |z_k|} < C$$

for all j and k. Such convergence is referred to as restricted nontangential convergence.

The subject matter, at least in several complex variables, remained dormant until the late 1960's when a series of papers by A. Koranyi [Ko2], E. M. Stein and N. J. Weiss [SW], and Weiss [We], extended the above results to bounded symmetric domains and Siegel domains of type II. For results concerning boundary behavior of Poisson integrals on symmetric spaces in general the reader is referred to the survey article of Koranyi [Ko4], and the more recent paper of Stein [Ste3].

7.1. Admissible Limits of Poisson Integrals.

The extension of Fatou's theorem to $\mathcal{M}$-harmonic functions on the ball is due to Koranyi [Ko2]. It turns out that the appropriate analogue for the ball of the nontangential approach regions are the **admissible domains** of Koranyi which are defined as follows:

Definition. *For $\zeta \in S$ and $\alpha > 1$, let*

$$(7.1) \qquad D_\alpha(\zeta) = \{z \in B : |1 - \langle z, w \rangle| < \frac{\alpha}{2}(1 - |z|^2)\}.$$

When $\alpha \le 1$, $D_\alpha(\zeta) = \phi$, and $\bigcup_{\alpha > 1} D_\alpha(\zeta) = B$ for every $\zeta \in S$. The domains D_α also satisfy

$$(7.2) \qquad U(D_\alpha(\zeta)) = D_\alpha(U\zeta)$$

for every $U \in \mathcal{U}$.

To get an idea of the shape of these approach regions, we take $\zeta = e_1$. Thus $z \in D_\alpha(e_1)$ if and only if

$$|1 - z_1| < \frac{\alpha}{2}(1 - |z|^2).$$

If we restrict ourselves to the complex line $\{\lambda e_1 : \lambda \in \mathbb{C}\}$ through 0 and e_1 we get the usual nontangential approach region

$$|1 - \lambda| < \frac{\alpha}{2}(1 - |\lambda|^2).$$

However, the intersection of $D_\alpha(e_1)$ with the $2n - 1$ dimensional real space obtained by setting $y_1 = 0$ gives

$$|1 - x_1| < \frac{\alpha}{2}(1 - x_1^2 - |z'|^2),$$

where $z' = (z_2, ..., z_n)$. With a little algebra, the above inequality is equivalent to

$$(x_1 - \tfrac{1}{\alpha})^2 + |z'|^2 < (1 - \tfrac{1}{\alpha})^2,$$

which is a $(2n - 1)$ dimensional ball centered at $(\frac{1}{\alpha}, 0, ..., 0)$ and tangent to S at e_1.

As in (5.10) and (5.11), if $f \in L^1(S)$ or ν is a complex measure on S, the Poisson-Szegö integral of f or ν is denoted by $\mathcal{P}[f]$ and $\mathcal{P}[\nu]$ respectively. Our goal in this chapter is to prove the following version of Fatou's theorem:

Theorem 7.1.

(a) If $F(z) = \mathcal{P}[f](z)$, where $f \in L^1(S)$, then for every $\alpha > 1$

$$(7.3) \qquad \lim_{\substack{z \to \zeta \\ z \in D_\alpha(\zeta)}} F(z) = f(\zeta) \qquad \sigma \text{ - a.e. on } S.$$

(b) If $F(z) = \mathcal{P}[\nu](z)$, where ν is a complex measure on S which is singular with respect to σ, then for every $\alpha > 1$,

$$(7.4) \qquad \lim_{\substack{z \to \zeta \\ z \in D_\alpha(\zeta)}} F(z) = 0 \qquad \sigma \text{ - a.e. on } S.$$

An immediate consequence of the above is the following: Suppose ν is a complex measure on S and $F(z) = \mathcal{P}[\nu](z)$. Let $d\nu = f \, d\sigma + d\nu_s$, where $f \in L^1(S)$ and ν_s is singular with respect to σ on S, be the Lebesgue decomposition of ν. Then for every $\alpha > 1$,

$$(7.5) \qquad \lim_{\substack{z \to \zeta \\ z \in D_\alpha(\zeta)}} F(z) = f(\zeta) \qquad \sigma \text{ - a.e. on } S.$$

7.2. Maximal Functions of Measures.

To consider boundary limits of Poisson integral of measures and functions, we introduce the following maximal functions on S:

Definition. Let ν be a complex measure on S. For $\zeta \in S$, define

$$(7.6) \qquad M[\nu](\zeta) = \sup_{\delta > 0} \frac{|\nu|(Q(\zeta, \delta))}{\delta^n},$$

where $|\nu|$ denotes the total variation of ν and $Q(\zeta, \delta)$ are the non-isotropic balls introduced in (5.5). Similarly, if $f \in L^1(S)$, set

$$(7.7) \qquad M[f](\zeta) = \sup_{\delta > 0} \frac{1}{\delta^n} \int_{Q(\zeta,\delta)} |f| \, d\sigma.$$

The term δ^n in the above definition of the maximal functions comes from the fact that $\sigma(Q(\zeta, \delta)) \approx \delta^n$, which we now prove.

Lemma 7.2. ([Ru3; Proposition 5.1.4]) *For $n \geq 1$, there exists a constant A_n such that for all δ, $0 \leq \delta \leq 2$,*

$$\left(\frac{\delta}{2}\right)^n \leq \sigma(Q(\zeta, \delta)) \leq A_n \delta^n.$$

Proof. Suppose $n > 1$. Since σ is $\mathcal{U}$-invariant, it suffices to take $\zeta = e_1$. Thus

$$Q(e_1, \delta) = Q_\delta = \{t \in S : |1 - t_1| < \delta\}.$$

Since χ_{Q_δ} is a function of t_1 only, by formula (1.10),

$$\sigma(Q_\delta) = \frac{n-1}{\pi} \int_{E(\delta)} (1 - |\lambda|^2)^{n-2} \, dA(\lambda),$$

where $E(\delta) = \{\lambda \in U : |1 - \lambda| < \delta\}$, and A is area measure on U. Consider the change of variables

$$1 - \lambda = \frac{\delta}{z}.$$

This maps $E(\delta)$ onto

$$E'(\delta) = \{z = x + iy : |z| > 1 \quad \text{and} \quad x > \frac{\delta}{2}\},$$

and by the change of variable formula (2.15) transforms the above integral into

$$\sigma(Q_\delta) = \frac{n-1}{\pi} \int_{E'(\delta)} \delta^n (2x - \delta)^{n-2} |z|^{-2n} \, dx \, dy,$$

or alternately,

$$\frac{\sigma(Q_\delta)}{\delta^n} = \frac{n-1}{\pi} \int_{E'(\delta)} (2x - \delta)^{n-2} |z|^{-2n} \, dx \, dy.$$

As δ decreases to 0, both $E'(\delta)$ and the integrand above increase. Therefore, by the monotone convergence theorem,

$$\lim_{\delta \to 0} \frac{\sigma(Q_\delta)}{\delta^n} = \frac{n-1}{\pi} \int_{E'(0)} 2^{n-2} x^{n-2} |z|^{-2n} \, dx \, dy$$

$$= \frac{2^{n-2}(n-1)}{\pi} \int_1^\infty \int_{-\frac{\pi}{2}}^{\frac{\pi}{2}} r^{-n-1} (\cos \theta)^{n-2} \, d\theta \, dr$$

$$= \frac{2^{n-2}(n-1)}{n\pi} \int_{-\frac{\pi}{2}}^{\frac{\pi}{2}} (\cos \theta)^{n-2} \, d\theta$$

$$= \frac{2^n}{4\sqrt{\pi}} \frac{\Gamma(\frac{n+1}{2})}{\Gamma(\frac{n}{2} + 1)}.$$

Thus $\sigma(Q_\delta) \leq A_n \delta^n$. On the otherhand, since $Q_2 = S$, we have for $\delta < 2$,

$$\frac{\sigma(Q_\delta)}{\delta^n} \geq \frac{\sigma(S)}{2^n} = \frac{1}{2^n},$$

which proves the result for $n > 1$. The case $n = 1$ is straightforward and consequently is omitted. $\square$

As was indicated in Section 5.1, the function

$$d(z,w) = |1 - \langle z, w \rangle|^{\frac{1}{2}}$$

satisfies the triangle inequality on $\overline{B}$, and defines a metric on S. Since this result is needed in the subsequent lemma, we include the proof.

Lemma 7.3. For a, b, $c \in \overline{B}$,

$$d(a,c) \leq d(a,b) + d(b,c).$$

Proof. Since d is $\mathcal{U}$ invariant, it suffices to take $b = re_1$, $0 \leq r \leq 1$. Thus we have to show that

$$|1 - \langle a, c \rangle| \leq \left(|1 - ra_1|^{\frac{1}{2}} + |1 - rc_1|^{\frac{1}{2}} \right)^2$$

Set $a' = a - a_1 e_1$ and $c' = c - c_1 e_1$. Then

$$|1 - \langle a, c \rangle| = |1 - a_1 \bar{c}_1 - \langle a', c' \rangle| \leq |1 - a_1 \bar{c}_1| + |a'||c'|.$$

But

$$\begin{aligned}|1 - a_1 \bar{c}_1| &= |1 - ra_1 + a_1(r - \bar{c}_1)| \\ &\leq |1 - ra_1| + |a_1||r - \bar{c}_1| \\ &\leq |1 - ra_1| + |1 - rc_1|,\end{aligned}$$

and

$$|a'|^2 \leq 1 - |a_1^2| \leq 1 - |ra_1|^2 \leq 2|1 - ra_1|.$$

Similarly, $|c'|^2 \leq 2|1 - rc_1|$. Combining the above gives

$$|1 - \langle a, c \rangle| \leq |1 - ra_1| + |1 - rc_1| + 2\sqrt{|1 - ra_1||1 - rc_1|}$$

$$= \left(|1 - ra_1|^{\frac{1}{2}} + |1 - rc_1|^{\frac{1}{2}} \right)^2. \quad \square$$

One of they key steps in proving the existence of admissible limits will be the following:

Theorem 7.4. *There exists a constant C_n, depending only on n, such that for any complex measure ν on S,*

$$(7.8) \qquad \sigma(\{\zeta \in S : M[\nu](\zeta) > t\}) \leq C_n \frac{\|\nu\|}{t} \qquad \text{for all} \quad t > 0.$$

For the proof of the above inequality we will need the following elementary covering lemma.

Lemma 7.5. ([Ru3]) *Suppose E is the union of a finite collection $\{Q(\zeta_i, \delta_i)\}$ of non-isotropic balls, then there exists a finite disjoint subcollection $\{Q(\zeta_{i_k}, \delta_{i_k})\}_{k=1}^m$ such that*

$$(7.9) \qquad\qquad E \subset \bigcup_{k=1}^m Q(\zeta_{i_k}, 9\delta_{i_k})$$

and

$$(7.10) \qquad\qquad \sigma(E) \le B_n \sum_{k=1}^m \sigma(Q(\zeta_{i_k}, \delta_{i_k})),$$

where B_n is a constant depending only on n.

Proof. Without loss of generality we can assume that the balls $\{Q(\zeta_i, \delta_i)\}$ have been ordered such that $\delta_i \ge \delta_{i+1}$. Set $Q_i = Q(\zeta_i, \delta_i)$.

Let $i_1 = 1$. Suppose $k \ge 1$ and i_k have been chosen. If Q_{i_k} intersects Q_i for every $i > i_k$, we stop the process. If not, let i_{k+1} be the first index such that $Q_{i_{k+1}}$ is disjoint from Q_{i_k}. Since the initial collection is finite, this process terminates at some step $k = m$.

We now show that the collection $\{Q_{i_k}\}$ has the desired properties. To every $i < i_m$, there corresponds some k such that

$$i_k \le i < i_{k+1}.$$

Suppose $t \in Q_i$ and $\zeta \in Q_i \cap Q_{i_k}$, which is nonempty by construction. Then

$$d(t, \zeta_{i_k}) \le d(t, \zeta_i) + d(\zeta_i, \zeta) + d(\zeta, \zeta_{i_k})$$
$$< \sqrt{\delta_i} + \sqrt{\delta_i} + \sqrt{\delta_{i_k}} \le 3\sqrt{\delta_{i_k}}.$$

If $i \ge i_m$, then $Q_i \cap Q_{k_m} \ne \phi$. Therefore, for every index i, $Q_i \subset Q(\zeta_{i_k}, 9\delta_{i_k})$ for some i_k, which proves (7.9). Thus if we set

$$B_n = \sup_{\delta > 0} \frac{\sigma(Q_{9\delta})}{\sigma(Q_\delta)},$$

which is finite by Lemma 7.2, we obtain (7.10). $\quad\square$

Proof of Theorem 7.4. Fix $t > 0$. We first note that for fixed $\delta > 0$, the function

$$\zeta \to \frac{|\nu|(Q(\zeta, \delta))}{\delta^n}$$

is a lower semicontinuous function on S. Thus $M[\nu](\zeta)$ is also lower semicontinuous. Therefore,

$$E_t = \{\zeta \in S : M[\nu](\zeta) > t\}$$

is an open subset of S. Let K be a compact subset of E_t. For each $\zeta \in K$, there exists a $\delta_\zeta > 0$ such that

$$|\nu|(Q(\zeta, \delta_\zeta)) > \delta_\zeta^n t.$$

By the compactness of K, there exists a finite subcollection $\{Q(\zeta_i, \delta_i)\}$ which also covers K. Thus by Lemma 7.5, there exists a finite disjoint subcollection $\{Q(\zeta_{i_k}, \delta_{i_k})\}$ satisfying

$$\sigma(K) \leq B_n \sum_k \sigma(Q((\zeta_{i_k}, \delta_{i_k}))) \leq B_n A_n \sum_k \delta_{i_k}^n$$

$$\leq \frac{A_n B_n}{t} \sum_k |\nu|(Q((\zeta_{i_k}, \delta_{i_k})))$$

$$= \frac{C_n}{t} |\nu|\left(\bigcup_k Q(\zeta_{i_k}, \delta_{i_k})\right) \leq \frac{C_n}{t} \|\nu\|.$$

The result now follows by inner regularity of the measure $|\nu|$. □

7.3. Differentiation Theorems.

We now use Theorem 7.4 to prove the following:

Theorem 7.6. *If $f \in L^1(S)$, then*

$$(7.11) \qquad \lim_{\delta \to 0^+} \frac{1}{\delta^n} \int_{Q(\zeta, \delta)} |f - f(\zeta)| \, d\sigma = 0$$

for almost every $\zeta \in S$.

Proof. If f is continuous, then (7.11) holds for every $\zeta \in S$. For arbitrary $f \in L^1(S)$ and $\epsilon > 0$, choose $g \in C(S)$ such that $\|f - g\|_1 < \epsilon$.

Let

$$T_f(\zeta) = \limsup_{\delta \to 0} \frac{1}{\delta^n} \int_{Q(\zeta, \delta)} |f - f(\zeta)| \, d\sigma.$$

Then

(a) $T_f(\zeta) \leq T_g(\zeta) + T_{f-g}(\zeta)$, and

(b) $T_{f-g}(\zeta) \leq A_n |f(\zeta) - g(\zeta)| + M[f - g](\zeta)$,

where A_n is the constant of Lemma 7.2. Since g is continuous, $T_g(\zeta) = 0$ for all $\zeta \in S$, and therefore,

$$T_f(\zeta) \leq A_n |f(\zeta) - g(\zeta)| + M[f - g](\zeta).$$

Thus for $t > 0$,

$$\sigma(\{\zeta : T_f(\zeta) > t\}) \le \sigma(\{\zeta : |f(\zeta) - g(\zeta)| > \frac{t}{2A_n}\})$$
$$+ \sigma(\{\zeta : M[f - g](\zeta) > \frac{t}{2}\}).$$

By Kolmogorov's inequality

$$\sigma(\{\zeta : |f(\zeta) - g(\zeta)| > \frac{t}{2A_n}\}) \le \frac{2A_n}{t}\|f - g\|_1 < \frac{2\epsilon A_n}{t},$$

and by Theorem 7.4,

$$\sigma(\{\zeta : M[f - g](\zeta) > \frac{t}{2}\}) \le \frac{2C_n}{t}\|f - g\|_1 < \frac{2\epsilon C_n}{t}.$$

Thus there exists a positive constant C, depending only on n, such that

$$\sigma(\{\zeta : T_f(\zeta) > t\}) < \frac{C\epsilon}{t}.$$

Since $\epsilon > 0$ was arbitrary, we have $\sigma(\{\zeta : T_f(\zeta) > t\}) = 0$ for every $t > 0$. Thus $T_f(\zeta) = 0$ a.e. on S. $\square$

In addition to the operator T_f, we also consider the upper derivate of a measure which is defined as follows:

Definition. *If ν is a finite positive measure on S, define the **upper derivate** $\overline{D}\nu$ of ν on S by*

(7.12) $$\overline{D}\nu(\zeta) = \limsup_{\delta \to 0+} \frac{\nu(Q(\zeta, \delta))}{\delta^n}.$$

If the limit in (7.12) exists, then we denote the value by $D\nu(\zeta)$. One of the advantages of considering the upper derivate of a measure is that unlike $M[\nu]$, it depends only on the behavior of $\nu(Q(\zeta, \delta))$ for small values of δ.

Lemma 7.7. *If ν is a finite positive singular measure on S, then*

$$D\nu = 0 \qquad \sigma \text{ - a.e. on } \quad S.$$

Proof. Suppose the conclusion is false. Then there exists a positive constant a, a compact set $K \subset S$ with $\sigma(K) > 0$, $\nu(K) = 0$, such that $\overline{D}\nu(\zeta) > a$ for all $\zeta \in K$.

Let $\epsilon > 0$ be arbitrary. Choose an open set $\Omega \subset S$ such that $K \subset \Omega$ and $\nu(\Omega) < \epsilon$. Since $\overline{D}\nu(\zeta) > a$ for every $\zeta \in K$, we can cover K by finitely many $Q(\zeta_i, \delta_i)$ such that $Q(\zeta_i, \delta_i) \subset \Omega$ and $\nu(Q(\zeta_i, \delta_i)) > a\,\delta_i^n$. By Lemma 7.5, there exists a finite disjoint subcollection $\{Q(\zeta_{i_k}, \delta_{i_k})\}$ such that

$$\sigma(K) \leq B_n \sum_k \sigma(Q(\zeta_{i_k}, \delta_{i_k}))$$

$$\leq \frac{C_n}{a} \sum_k \nu(Q(\zeta_{i_k}, \delta_{i_k})) \leq \frac{C_n}{a}\,\nu(\Omega) < \frac{C_n}{a}\,\epsilon.$$

Since $\epsilon > 0$ was arbitrary, we have $\sigma(K) = 0$, which is a contradiction. $\quad\square$

7.4. The Admissible Maximal Function.

Definition. *Let* $F \in C(B)$. *For* $\alpha > 1$, *define the* **admissible maximal function** $M_\alpha F$ *on* S *by*

$$(7.13) \qquad (M_\alpha F)(\zeta) = \sup\{|F(z)| : z \in D_\alpha(\zeta)\}.$$

Since F is continuous, $\{\zeta \in S : (M_\alpha F)(\zeta) \leq t\}$ is closed in S for every real number t. Thus $M_\alpha F$ is lower semicontinuous.

We now come to the main result needed in the proof of Theorem 7.1.

Theorem 7.8. *([Ko2]) For every* $\alpha > 1$, *there exists a finite constant* $A(\alpha)$ *such that*

$$(7.14) \qquad M_\alpha \mathcal{P}[\nu] \leq A(\alpha)\,M[\nu]$$

for every complex measure ν *on* S.

Proof. Since $|\mathcal{P}[\nu]| \leq \mathcal{P}[|\nu|]$, it suffices to consider the case where ν is a finite positive measure on S.

Fix a $\zeta \in S$ for which $M[\nu](\zeta) < \infty$, and fix $z \in D_\alpha(\zeta)$. Let $r = |z|$ and set $\delta = \alpha(1 - r)$. Let N be the smallest integer such that $2^N\delta > 2$. Set

$$V_0 = \{\omega \in S : |1 - \langle \omega, \zeta \rangle| < \delta\},$$

and for $k = 1, 2, ..., N$, set

$$V_k = \{\omega \in S : 2^{k-1}\delta \leq |1 - \langle \omega, \zeta \rangle| < 2^k\delta\}.$$

Then

$$\mathcal{P}[\nu](z) = \int_S \mathcal{P}(z, \omega)\,d\nu(\omega)$$

$$= \int_{V_0} \mathcal{P}(z, \omega)\,d\nu(\omega) + \sum_{k=1}^N \int_{V_k} \mathcal{P}(z, \omega)\,d\nu(\omega).$$

Suppose $\omega \in V_0$. Then

$$\mathcal{P}(z, \omega) = \frac{(1 - |z|^2)^n}{|1 - \langle z, \omega \rangle|^{2n}} \leq \frac{2^n}{(1 - |z|)^n} = \frac{2^n \alpha^n}{\delta^n}.$$

Therefore,

(7.15) $$\int_{V_0} \mathcal{P}(z, \omega) \, d\nu(\omega) \leq \frac{2^n \alpha^n}{\delta^n} \nu(V_0) \leq 2^n \alpha^n \, M[\nu](\zeta).$$

In the above we have used the fact that

$$\nu(V_0) \leq M[\nu](\zeta) \delta^n.$$

To estimate the integral over V_k, we first need to establish the following inequality for the Poisson kernel: for $z \in D_\alpha(\zeta)$, $\omega \in S$,

(7.16) $$|1 - \langle \zeta, \omega \rangle| < 4\alpha \, |1 - \langle z, \omega \rangle|,$$

and thus

(7.17) $$\mathcal{P}(z, \omega) \leq \frac{(16\alpha^2)^n (1 - r^2)^n}{|1 - \langle \zeta, \omega \rangle|^{2n}} \leq \frac{(32\alpha^2)^n (1 - r)^n}{|1 - \langle \zeta, \omega \rangle|^{2n}}.$$

To prove (7.16), we first note that since $z \in D_\alpha(\zeta)$,

$$|1 - \langle z, \zeta \rangle| < \alpha(1 - |z|) \leq \alpha \, |1 - \langle z, \omega \rangle|.$$

Thus by the triangle inequality,

$$|1 - \langle \zeta, \omega \rangle|^{\frac{1}{2}} \leq |1 - \langle z, \zeta \rangle|^{\frac{1}{2}} + |1 - \langle z, \omega \rangle|^{\frac{1}{2}}$$
$$< (1 + \sqrt{\alpha})|1 - \langle z, \omega \rangle|^{\frac{1}{2}} < 2\sqrt{\alpha}|1 - \langle z, \omega \rangle|^{\frac{1}{2}},$$

which proves (7.16). Inequality (7.17) follows immediately from the formula for the Poisson-Szegö kernel.

Suppose $\omega \in V_k$, $k = 1, 2, \ldots$. Then by (7.17),

$$\mathcal{P}(z, \omega) \leq \frac{(32)^n \alpha^{2n} (1 - r)^n}{2^{2n(k-1)} \delta^{2n}} = \frac{(32\alpha)^n 2^{-2n(k-1)}}{\delta^n}.$$

Therefore, since $\nu(V_k) \leq M[\nu](\zeta)(2^k \delta)^n$,

(7.18) $$\int_{V_k} \mathcal{P}(z, \omega) \, d\nu(\omega) \leq (32)^n \alpha^n M[\nu](\zeta) \, 2^{2n} 2^{-nk}.$$

Combining (7.15) and (7.18) and summing over k gives the desired result. $\quad\square$

Lemma 7.9. *If ν is a finite positive measure on S satisfying $D\nu(\zeta) = 0$, $\zeta \in S$, then*

$$(7.19) \qquad \lim_{\substack{z \to \zeta \\ z \in D_\alpha(\zeta)}} \mathcal{P}[\nu](z) = 0.$$

Proof. Let $\epsilon > 0$ be arbitrary. Since $D\nu(\zeta) = 0$, there exists $\delta_o > 0$ such that

$$(7.20) \qquad \nu(Q(\zeta, \delta)) < \epsilon \delta^n$$

for all δ, $0 < \delta < \delta_o$. Let $Q_o = Q(\zeta, \delta_o)$. Set $\nu_o = \nu_{|Q_o}$, and $\nu_1 = \nu - \nu_o$. Then

$$\mathcal{P}[\nu_1](z) = \int_{S \sim Q_o} \mathcal{P}(z, w) \, d\nu(w).$$

If $w \in S \sim Q_o$, then $|1 - \langle w, \zeta \rangle| \geq \delta_o$. Thus by inequality (7.17), for $z \in D_\alpha(\zeta)$,

$$\mathcal{P}[\nu_1](z) \leq \frac{(16\alpha^2)^n}{\delta_o^{2n}} (1 - |z|^2)^n.$$

Thus

$$\lim_{\substack{z \to \zeta \\ z \in D_\alpha(\zeta)}} \mathcal{P}[\nu_1](z) = 0.$$

Also, by (7.20), $M[\nu_o](\zeta) < \epsilon$. Therefore, by Theorem 7.8,

$$\limsup_{\substack{z \to \zeta \\ z \in D_\alpha(\zeta)}} \mathcal{P}[\nu_o](z) \leq A(\alpha) \, \epsilon,$$

from which (7.19) now follows. $\square$

Proof of Theorem 7.1. Suppose ν is a complex singular measure on S. Since $|\nu|$ is also singular, by Lemma 7.7 $D|\nu|(\zeta) = 0$ σ- a.e. on S. Thus by Lemma 7.9,

$$\lim_{\substack{z \to \zeta \\ z \in D_\alpha(\zeta)}} \mathcal{P}[|\nu|](z) = 0 \qquad \sigma \text{ - a.e. on } S.$$

Since $|\mathcal{P}[\nu]| \leq \mathcal{P}[|\nu|]$, this proves Theorem 7.1 (b).

If $f \in L^1(S)$, then by Theorem 7.6,

$$\lim_{\delta \to 0+} \frac{1}{\delta^n} \int_{Q(\zeta, \delta)} |f - f(\zeta)| \, d\sigma = 0$$

for almost every $\zeta \in S$. Fix such a ζ and define μ on S by

$$\mu(E) = \int_E |f - f(\zeta)| \, d\sigma,$$

for all Borel subsets E of S. Then

$$D\mu(\zeta) = 0,$$

and

$$|\mathcal{P}[f](z) - f(\zeta)| \le \mathcal{P}[\mu](z).$$

Therefore by Lemma 7.9 again,

$$\lim_{\substack{z \to \zeta \\ z \in D_\alpha(\zeta)}} |\mathcal{P}[f](z) - f(\zeta)| \le \lim_{\substack{z \to \zeta \\ z \in D_\alpha(\zeta)}} \mathcal{P}[\mu](z) = 0,$$

which proves Theorem 7.1 (a). $\square$

7.5. Weighted Radial Limits of Poisson Integrals.

Suppose ν is a complex measure on S, and $F = \mathcal{P}[\nu]$. By Theorem 7.1 (a), the absolutely continuous part of ν is recovered as the admissible limit of F. We now show how the discrete part of ν can be recovered as a weighted radial limit of F.

Theorem 7.10. *Let ν be a complex measure on S. Then*

$$(7.21) \qquad \lim_{r \to 1} (1 - r)^n \mathcal{P}[\nu](r\zeta) = 2^n \nu(\{\zeta\}) \qquad \text{for every } \zeta \in S.$$

Proof. By linearity, we can without loss of generality assume that $\nu \ge 0$.

Fix $\zeta \in S$, and suppose $\nu(\{\zeta\}) = 0$. Let $\epsilon > 0$ be arbitrary. Then there exists a $\delta > 0$ such that $\nu(Q(\zeta, \delta)) < \epsilon$. Fix such a δ_o and set $Q_o = Q(\zeta, \delta_o)$. Let $\nu_o = \nu|_{Q_o}$, and $\nu_1 = \nu - \nu_o$. Then $D\nu_1(\zeta) = 0$, and thus by Lemma 7.9,

$$\lim_{r \to 1} \mathcal{P}[\nu_1](r\zeta) = 0.$$

On the other hand, if $\omega \in Q(\zeta, \delta_o)$,

$$(1 - r)^n \mathcal{P}(r\zeta, \omega) \le 2^n.$$

Therefore,

$$\limsup_{r \to 1} (1 - r)^n \mathcal{P}[\nu_o](r\zeta) \le 2^n \nu(Q_o) < 2^n \epsilon.$$

Combining the two gives

$$\limsup_{r \to 1} (1 - r)^n \mathcal{P}[\nu](r\zeta) \le 2^n \nu(Q_o) < 2^n \epsilon.$$

Since $\epsilon > 0$ was arbitrary, we have

$$\lim_{r \to 1} (1 - r)^n \mathcal{P}[\nu](r\zeta) = 0.$$

Suppose $\nu(\{\zeta\}) = a > 0$. Set $\mu = \nu - a\,\delta_\zeta$, where δ_ζ is pointmass at ζ. Then $\mu(\{\zeta\}) = 0$, and hence

$$\lim_{r \to 1} (1 - r)^n \mathcal{P}[\mu](r\zeta) = 0.$$

But

$$\mathcal{P}[\mu](r\zeta) = \mathcal{P}[\nu](r\zeta) - a\mathcal{P}(r\zeta, \zeta)$$
$$= \mathcal{P}[\nu](r\zeta) - a\frac{(1 - r^2)^n}{(1 - r)^{2n}},$$

from which the result now follows. $\square$

When $n > 1$, Theorem 7.10 can be generalized to allow more than just radial convergence.

Definition. *Let $\zeta \in S$. A sequence $\{z^j\}$ in B with $z^j \to \zeta$ is said to converge* **hyporadially** *to ζ if*

(a) for each j, there exists r_j, $0 < r_j < 1$, such that $z^j - r_j\zeta \perp \zeta$, and

(b) $\displaystyle \lim_{j \to \infty} \frac{|z^j - r_j\zeta|^2}{(1 - r_j)} = 0$.

The term hyporadially is adapted from the term **hypoadmissibly** which is the appropriate setting for the generalization of the classical Lindelöf principle to the ball in $\mathbb{C}^n$ [Kr1, Theorem 8.7.3]. A sequence $\{z^j\}$ in B converges hypoadmissibly to $\zeta \in S$ if there is an $\alpha > 1$ such that $z^j \in D_\alpha(\zeta)$ for all j, and such that

$$\lim_{j \to \infty} \frac{|z^j - \langle z^j, \zeta \rangle \zeta|^2}{|1 - \langle z^j, \zeta \rangle|} = 0.$$

The hypothesis (a) implies that for every j,

$$z^j = r_j\,\zeta + w^j,$$

where $w^j \perp \zeta$, i.e., $\langle z^j, \zeta \rangle = r_j\zeta$ for some r_j, $0 < r_j < 1$. When $n = 1$, this just means that $z_j = r_j\zeta$. With the above decomposition we have

$$z^j - \langle z^j, \zeta \rangle \zeta = z^j - r_j\zeta = w^j,$$

and

$$1 - \langle z^j, \zeta \rangle = 1 - r_j.$$

The important fact about hyporadial convergence is that if $z^j \to \zeta$ hyporadially, then

$$\lim_{j \to \infty} (1 - |z^j|)^n \, \mathcal{P}(z^j, \zeta) = 2^n.$$

With this fact, the same proof as of Theorem 7.10 gives the following:

Theorem 7.11. *Let ν be a complex measure on S. Then for every sequence $\{z^j\}$ in B which converges hyporadially to $\zeta \in S$,*

$$(7.22) \qquad \lim_{j \to \infty} (1 - |z^j|)^n \mathcal{P}[\nu](z^j) = 2^n \, \nu(\{\zeta\}).$$

7.6. Remarks.

We close this chapter with several comments concerning related results in the ball and also the polydisc.

(1) Using the results of this section, one can easily show that an alternate version of Fatou's theorem is as follows: if ν is a finite complex measure on S, then

$$\lim_{\substack{z \to \zeta \\ z \in D_\alpha(\zeta)}} \mathcal{P}[\nu](z) = \mathcal{D}\nu(\zeta), \qquad \text{for a.e.} \quad \zeta \in S,$$

where $\mathcal{D}\nu$ is the symmetric derivative of ν defined by

$$\mathcal{D}\nu(\zeta) = \lim_{\delta \to 0+} \frac{\nu(Q(\zeta, \delta))}{\sigma(Q(\zeta, \delta))},$$

whenever the limit exists. When $n = 1$, it is known that

$$\lim_{r \to 1} \mathcal{P}[\nu](re^{i\theta}) = \mathcal{D}\nu(e_{i\theta})$$

at every $e^{i\theta}$ where $\mathcal{D}\nu(e^{i\theta})$ exists. When $n > 1$, this fails in general ([Ru3, Examples 5.4.13]). However, for positive pluriharmonic functions, we still have the following:

Theorem 7.12. *([RaU]) Let u be a positive pluriharmonic function on B_n, $n > 1$ with boundary measure ν. Let $\zeta_o \in S$, and suppose $0 \leq L \leq \infty$. Then*

$$\mathcal{D}\nu(\zeta_o) = L \qquad \text{if and only if} \qquad \lim_{r \to 1} u(r\zeta_o) = L.$$

(2) Our second remark deals with the area integral and the local Fatou theorem. For f $\mathcal{M}$-harmonic on B, and $\alpha > 1$, the square area integral of f is the function $S_\alpha f$ on S defined by

$$(7.23) \qquad S_\alpha f(\zeta) = \left(\int_{D_\alpha(\zeta)} |\tilde{\nabla} f|^2 \, d\lambda \right)^{\frac{1}{2}}, \qquad \zeta \in S.$$

The following theorem, which we state without proof, provides equivalent conditions for the existence of admissible limits.

Theorem 7.13. *([Ge], [Pu]) Let f be $\mathcal{M}$-harmonic on B, and let E be a measurable subset of S. Then the following are equivalent.*

(a) $M_\alpha f(\zeta) < \infty$ *for almost every* $\zeta \in E$.

(b) $S_\alpha f(\zeta) < \infty$ *for almost every* $\zeta \in E$.

(c) f *has admissible limits at almost every* $\zeta \in E$.

The theorem was first stated by R. Putz in [Pu], although, as was indicated by D. Geller [Ge], there appears to be an error in part of the proof. The analogous theorem for holomorphic functions on strictly pseudoconvex domains was stated and proved by E. M. Stein [Ste2].

In connection with the area integral, we note that if $f \in \mathcal{H}^2(B)$, then by Theorem 6.18,

$$\int_B (1 - |w|^2)^n |\widetilde{\nabla} f(w)|^2 \, d\lambda(w) < \infty.$$

As we will see in the next chapter (Proposition 8.13), this will be the case if and only if $S_\alpha f \in L^2(S)$.

(3) Our final comment concerns Fatou's theorem for weakly harmonic functions in U^2. Our notation will be the same as in Remark (3) of Section 5.5, except that we set

$$\lambda_1 = R \cos \theta, \quad \lambda_2 = R \sin \theta, \qquad R > 0, \quad \theta \in [0, \tfrac{\pi}{2}].$$

Then by Theorem 5.12 every positive solution of $\widetilde{\Delta} F = 2(R^2 - \tfrac{1}{2}) F$ is given by

$$F(z) = \mathcal{P}^\lambda[\nu](z) = \int_{T^2 \times [0, \frac{\pi}{2}]} \mathcal{P}^\lambda(z, t) \, d\nu(t, \theta)$$

where ν is a positive measure on $T^2 \times [0, \tfrac{\pi}{2}]$. If $f \in L^1(T^2 \times [0, \tfrac{\pi}{2}])$, set $\mathcal{P}^\lambda[f] = \mathcal{P}^\lambda[\nu]$, where ν is the measure given by $d\nu(t, \theta) = f(t, \theta) d\sigma(t) d\theta$, and σ is normalized Lebesgue measure on T^2.

For $R > 0$, a boundary point $(t, \theta) \in T^2 \times (0, \tfrac{\pi}{2})$ has to be thought of as a limit point of the geodesic curve $\Gamma_{t,\theta}$ defined by

$$(z_1, z_2) = (t_1 \tanh(sR \cos \theta), t_2 \tanh(sR \sin \theta)), \qquad s \geq 0.$$

If $\theta = \pi/4$ so that $\lambda_1 = \lambda_2$, then this equivalent to radial convergence to t. The result of P. Sjögrin [Sj] is as follows:

Theorem 7.14. *Let $R > 0$. If $f \in L^1(T^2 \times [0, \tfrac{\pi}{2}])$, then*

$$\frac{\mathcal{P}^\lambda[f](z)}{\mathcal{P}^\lambda[1](z)} \longrightarrow f$$

for a.e. $(t, \theta) \in T^2 \times (0, \tfrac{\pi}{2})$ *as* (z_1, z_2) *approaches* T^2 *along* $\Gamma_{t,\theta}$ *or within bounded hyberbolic distance of* $\Gamma_{t,\theta}$.

8.
Radial and Admissible
Boundary Limits of Potentials

When $n = 1$, the classical theorem of Littlewood [Li] states that if f is a subharmonic function in U satisfying

$$\sup_{0<r<1} \int_0^{2\pi} f^+(re^{i\theta})\, d\theta < \infty,$$

then

$$\lim_{r \to 1} f(rt) \quad \text{exists for a.e.} \quad t \in T.$$

In this chapter we will prove the extension of Littlewood's theorem to $\mathcal{M}$-subharmonic functions in B, and also investigate the existence of admissible limits.

8.1. Radial Limits of Potentials.

Suppose f is $\mathcal{M}$-subharmonic in B satisfying

$$(8.1) \qquad \sup_{0<r<1} \int_S f^+(rt)\, d\sigma(t) < \infty.$$

Then by the Riesz decomposition theorem

$$f(z) = \mathcal{P}[\nu_f](z) - G_{\mu_f}(z),$$

where μ_f is the Riesz measure of f which satisfies

$$(8.2) \qquad \int_B (1 - |w|^2)^n\, d\mu_f(w) < \infty$$

and ν_f is the boundary measure of f, i.e. $\mathcal{P}[\nu_f]$ is the least $\mathcal{M}$-harmonic majorant of f. If $d\nu_f = \hat{f}\, d\sigma + d\nu_s$, where ν_s is singular and $\hat{f} \in L^1(S)$, is the Lebesgue decomposition of ν_f, then by Theorem 7.1,

$$\lim_{\substack{z \to \zeta \\ z \in D_\alpha(\zeta)}} \mathcal{P}[\nu_f](z) = \hat{f}(\zeta) \qquad \text{a.e. on} \quad S.$$

Thus to investigate the boundary behavior of f, we need to consider the invariant potential G_{μ_f}. The theorem we will prove is as follows:

Theorem 8.1. *([Ul1,Ul2]) Let μ be a nonnegative regular Borel measure on B satisfying (8.2). Then*

$$\lim_{r \to 1} G_\mu(r\zeta) = 0 \qquad \text{for a.e.} \quad \zeta \in S.$$

As a consequence of Theorems 7.1 and 8.1, if f is $\mathcal{M}$-subharmonic on B satisfying (8.1), then

$$\lim_{r \to 1} f(rt) = \hat{f}(t) \qquad \text{for a.e.} \quad t \in S.$$

To prove Theorem 8.1, we consider two auxilliary functions which we now define. As in (4.2), for $z \in B$, let

$$E(z) = \{w \in B : |\varphi_z(w)| < \tfrac{1}{2}\}.$$

Define the functions V_1 and V_2 on B as follows:

(8.3) $$V_1(z) = \int_{E(z)} G(z,w)\,d\mu(w),$$

(8.4) $$V_2(z) = \int_{B \sim E(z)} G(z,w)\,d\mu(w).$$

Then $G_\mu(z) = V_1(z) + V_2(z)$.

Our first result will be to prove that the function V_2 has admissible limits a.e. on S. Thus, the failure of the existence of admissible limits in general lies with the function V_1.

Proposition 8.2. *Let μ be a nonnegative regular Borel measure on B satisfying (8.2) and let V_2 be defined by (8.4). Then for every $\alpha > 1$,*

$$\lim_{\substack{z \to \zeta \\ z \in D_\alpha(\zeta)}} V_2(z) = 0 \qquad \text{for a.e.} \quad \zeta \in S.$$

Proof. If $w \in B \sim E(z)$, then by inequality (6.5b) and identity (1.16),

(8.5) $$G(z,w) \le C\,(1 - |\varphi_z(w)|^2)^n = C\,\frac{(1 - |z|^2)^n (1 - |w|^2)^n}{|1 - \langle z,w \rangle|^{2n}},$$

where C is a constant depending only on n.

Fix an R, $0 < R < 1$, and let

$$B_R = \{z : |z| \le R\} \qquad \text{and} \qquad A_R = \{z : R < |z| < 1\}.$$

Then for all $z \in B$,

$$(8.6) \qquad V_2(z) \le C_R (1 - |z|^2)^n + C \int_{A_R} \frac{(1 - |z|^2)^n (1 - |w|^2)^n}{|1 - \langle z, w \rangle|^{2n}} \, d\mu(w).$$

where C_R is a constant depending on R. In the above we have used the fact that for $w \in B_R$, $|1 - \langle z, w \rangle| \ge (1 - R)$, and thus

$$\int_{B_R} \frac{(1 - |z|^2)^n (1 - |w|^2)^n}{|1 - \langle z, w \rangle|^{2n}} \, d\mu(w) \le \frac{(1 - |z|^2)^n}{(1 - R)^{2n}} \int_{B_R} (1 - |w|^2)^n \, d\mu(w)$$

$$= C_R (1 - |z|^2)^n.$$

Define a measure ν_R on S as follows: for $h \in C(S)$,

$$(8.7) \qquad \int_S h(t) \, d\nu_R(t) = \int_{A_R} h(\tfrac{w}{|w|})(1 - |w|^2)^n \, d\mu(w).$$

Then ν_R is a finite Borel measure on S. Thus since $|1 - \langle z, w \rangle| \ge \frac{1}{2}|1 - \langle z, \tfrac{w}{|w|} \rangle|$,

$$\int_{A_R} \frac{(1 - |z|^2)^n (1 - |w|^2)^n}{|1 - \langle z, w \rangle|^{2n}} \, d\mu(w)$$

$$\le 2^{2n} C \int_{A_R} \frac{(1 - |z|^2)^n}{|1 - \langle z, \tfrac{w}{|w|} \rangle|^{2n}} (1 - |w|^2) \, d\mu(w)$$

$$= C_n \, \mathcal{P}[\nu_R](z),$$

where C_n is a fixed constant independent of R. Therefore

$$(8.8) \qquad V_2(z) \le C_R (1 - |z|^2)^n + C_n \, \mathcal{P}[\nu_R](z).$$

Let $\alpha > 1$. Then by the above, for any R, $0 < R < 1$,

$$\limsup_{\substack{z \to \zeta \\ z \in D_\alpha(\zeta)}} V_2(z) \le \lim_{z \to \zeta} C_R (1 - |z|^2)^n + \limsup_{\substack{z \to \zeta \\ z \in D_\alpha(\zeta)}} C_n \, \mathcal{P}[\nu_R](z)$$

$$\le \sup_{z \in D_\alpha(\zeta)} C_n \, \mathcal{P}[\nu_R](z)$$

$$\le C_n A(\alpha) \, M[\nu_R](\zeta).$$

In the last inequality we have used Theorem 7.8. Now, by Theorem 7.4, there exists a constant A_n, depending only on n, such that

$$\sigma(\{\zeta \in S : M[\nu_R](\zeta) > \delta\}) \le \frac{A_n}{\delta} \|\nu_R\|$$

for every $\delta > 0$. Thus for any R, $0 < R < 1$,

$$\sigma(\{\limsup_{\substack{z \to \zeta \\ z \in D_\alpha(\zeta)}} V_2(z) > \delta\}) \leq \sigma(\{\zeta \in S : C_n A(\alpha) M[\nu_R](\zeta) > \delta\})$$

$$\leq \frac{A_n C_n A(\alpha)}{\delta} \|\nu_R\|.$$

Finally, since

$$\lim_{R \to 1} \|\nu_R\| = \lim_{R \to 1} \int_{A_R} (1 - |w|^2)^n \, d\mu(w) = 0,$$

we obtain that

$$\sigma(\{\limsup_{\substack{z \to \zeta \\ z \in D_\alpha(\zeta)}} V_2(z) > \delta\}) = 0$$

for every $\delta > 0$. Thus

$$\lim_{\substack{z \to \zeta \\ z \in D_\alpha(\zeta)}} V_2(z) = 0$$

for almost every $\zeta \in S$.

To complete the proof of Theorem 8.1, our next step will be to prove the following:

Proposition 8.3. *Let μ be a nonnegative regular Borel measure on B satisfying (8.2), and let V_1 be defined by (8.3). Then*

$$\lim_{r \to 1} V_1(r\zeta) = 0 \qquad \text{for a.e.} \quad \zeta \in S.$$

Once Proposition 8.3 has been proved, Theorem 8.1 then follows. To prove Proposition 8.3 we will need some preliminary results. The first of these is of independent interest, and is a generalization of a well know fact in the unit disc.

Lemma 8.4. $\rho(z,w) = |\varphi_z(w)|$ *defines a metric on B.*

Proof. Clearly, the only thing that needs to be proved is that ρ satisfies the triangle inequality, i.e.,

$$|\varphi_z(w)| \leq |\varphi_\zeta(z)| + |\varphi_\zeta(w)|$$

for all z, w, $\zeta \in B$. We first note that for $x, y \in [0,1]$,

$$1 - (x + y)^2 \leq \frac{(1 - x^2)(1 - y^2)}{(1 + xy)^2}.$$

Thus

$$1 - (|\varphi_\zeta(z)| + |\varphi_\zeta(w)|)^2 \leq \frac{(1 - |\varphi_\zeta(z)|^2)(1 - |\varphi_\zeta(w)|^2)}{(1 + |\varphi_\zeta(z)||\varphi_\zeta(w)|)^2}$$

$$\leq \frac{(1 - |\varphi_\zeta(z)|^2)(1 - |\varphi_\zeta(w)|^2)}{|1 - \langle \varphi_\zeta(z), \varphi_\zeta(w) \rangle|^2},$$

which by identities (1.15) and (1.16)

$$= \frac{(1 - |z|^2)(1 - |w|^2)}{|1 - \langle z, w \rangle|^2} = 1 - |\varphi_z(w)|^2,$$

from which the result now follows. □

For $z \in B$, $0 < r < 1$, let

(8.9) $$E_r(z) = \{w \in B : |\varphi_z(w)| < r\} = \varphi_z(B_r).$$

If $r \geq 1$, we set $E_r(z) = B$. We now list some basic properties of these sets which will be needed in the sequel.

Lemma 8.5.

(a) For $z, a \in B$, $0 < r < 1$,
$$\varphi_a(E_r(z)) = E_r(\varphi_a(z)).$$

(b) If $E_r(z) \cap E_{r'}(z') \neq \phi$ with $r' \leq r$, then $E_{r'}(z') \subset E_{3r}(z)$.

(c) For all $w \in E(z)$,
$$\tfrac{1}{3}(1 - |z|^2) \leq (1 - |w|^2) \leq 3(1 - |z|^2).$$

(d) For $0 < R < 1$, $E(z) \subset A_R$ for all z, $|z|^2 \geq 1 - \tfrac{1}{3}(1 - R^2)$.

Proof. The proof of (a) follows from the fact that $|\varphi_z(\varphi_a(w))| = |\varphi_{\varphi_a(z)}(w)|$ (see Lemma 4.5), and the proof of (b) is an immediate consequence of the fact that $\rho(z, w) = |\varphi_z(w)|$ satisfies the triangle inequality.

To prove (c) we first note that $w \in E(z)$ if and only if $w = \varphi_z(\zeta)$ with $|\zeta| < \tfrac{1}{2}$. Thus

$$1 - |w|^2 = \frac{(1 - |z|^2)(1 - |\zeta|^2)}{|1 - \langle z, \zeta \rangle|^2}$$

$$\leq (1 - |z|^2)\left(\frac{1 + |\zeta|}{1 - |\zeta|}\right) \leq 3(1 - |z|^2).$$

Since $w \in E(z)$ if and only if $z \in E(w)$, we get the reverse inequality by symmetry.

Finally, for the proof of (d), if $|z|^2 \geq 1 - \tfrac{1}{3}(1 - R^2)$, then $3(1 - |z|^2) \leq (1 - R^2)$. Thus by (c), if $w \in E(z)$, $(1 - |w|^2) \leq (1 - R^2)$, i.e., $|w| \geq R$. □

Let $\Pi : B \sim \{0\} \to S$ by $\Pi(z) = z/|z|$, and for $\zeta \in S$, $r, R \in (0, 1)$, set

(8.10) $$V_r^R(\zeta) = \Pi(E_r(R\zeta))$$

Lemma 8.6. *For $\zeta \in S, 0 < r \leq \frac{1}{2}, \frac{3}{4} < R < 1$,*

$$\sigma(V_r^R(\zeta)) \approx (1 - R^2)^n r^{2n-1}.$$

Proof. By unitary invariance, we may assume $\zeta = e_1$. For $0 < r \leq \frac{1}{2}$ and ρ small, let

$$\Omega_r^\rho = \{te^{i\theta} : 0 < 1 - t < \rho^2 r^2, |\theta| < \rho^2 r\}$$

and

$$N_r^\rho = \{\zeta \in S : \zeta_1 \in \Omega_r^\rho\}.$$

We show below that there exist positive constants c, c' independent of r and R such that for $\rho^2 = 1 - R^2$,

(8.11) $$N_{cr}^\rho \subset V_r^R(e_1) \subset N_{c'r}^\rho.$$

Assumming that (8.11) holds, we then obtain that $\sigma(V_r^R(e_1)) \approx \sigma(N_r^\rho)$. For $n = 1$, the result is now straight forward. For $n > 1$, by integral formula (1.10),

$$\sigma(N_r^\rho) = \frac{n-1}{\pi} \int_{|\theta| < \rho^2 r} \int_{1-\rho^2 r^2}^1 (1 - t^2)^{n-2} t \, dt d\theta$$

$$= \frac{1}{\pi} \rho^{2n} r^{2n-1},$$

from which the lemma now follows.

To prove (8.11), we first show that for c appropriately chosen, if $\zeta \in N_{cr}^\rho$, then $R\zeta \in E_r(Re_1)$. Thus $\zeta \in V_r^R(e_1)$. By identity (1.16), $R\zeta \in E_r(Re_1)$ if and only if

$$\frac{(1 - R^2)^2}{|1 - R^2 \zeta_1|^2} = 1 - |\varphi_{Re_1}(R\zeta)|^2 > 1 - r^2,$$

or equivalently,

$$|1 - R^2 \zeta_1|^2 < \frac{(1 - R^2)^2}{(1 - r^2)}.$$

If $\zeta \in N_{cr}^\rho$, then $\zeta_1 = te^{i\theta} \in \Omega_{cr}^\rho$. Thus

$$|1 - R^2 \zeta_1|^2 = |1 - R^2 te^{i\theta}|^2$$

$$= 2R^2 t(1 - \cos\theta) + (1 - R^2 t)^2$$

$$< \theta^2 + (R^2(1 - t) + (1 - R^2))^2.$$

In the above we have used the fact that $R^2 t < 1$ and that

$$2(1 - \cos\theta) = 4\sin^2 \tfrac{\theta}{2} < \theta^2.$$

Since $\zeta \in N_{cr}^\rho$ with $\rho^2 = (1 - R^2)$,

$$|\theta| < c(1 - R^2)r \qquad \text{and} \qquad 1 - t < c^2 r^2 (1 - R^2).$$

Therefore

$$|1 - R^2 \zeta_1|^2 < c^2 r^2 (1 - R^2)^2 + (c^2 r^2 R^2 (1 - R^2) + (1 - R^2))^2$$
$$\leq (1 - R^2)^2 (1 + 3c^2 r^2 + c^4 r^4).$$

Choose $c > 0$ such that $3c^2 < 1$. Then

$$1 + 3c^2 r^2 + c^4 r^4 < 1 + r^2 + (r^2)^2 < \frac{1}{1 - r^2}.$$

For this choice of c we have $N_{cr}^\rho \subset V_r^R(e_1)$ for all r, $0 < r \leq \tfrac{1}{2}$.

We now show that $V_r^R(e_1) \subset N_{c'r}^\rho$ for some $c' > 0$. Suppose $\zeta = (\zeta_1, \zeta') \in V_r^R(e_1)$. Then $\zeta = z/|z|$ for some $z \in E_r(Re_1)$. By (4.6), $z \in E_r(Re_1)$ if and only if

(8.12)
$$\frac{|z_1 - a|^2}{r^2 \beta^2} + \frac{|z'|^2}{r^2 \beta} < 1.$$

where

$$a = \frac{(1 - r^2) R}{1 - r^2 R^2} \qquad \text{and} \qquad \beta = \frac{1 - R^2}{1 - r^2 R^2}.$$

From the above, we have for $0 < r \leq \tfrac{1}{2}$,

$$|z'|^2 < r^2 \left(\frac{1 - R^2}{1 - r^2 R^2} \right) < c_1 r^2 (1 - R^2),$$

where c_1 is independent of r. Also, by (8.12),

$$|z_1 - R| \leq |z_1 - a| + |a - R|$$
$$\leq \frac{r(1 - R^2)}{1 - r^2 R^2} + \frac{Rr^2 (1 - R^2)}{1 - r^2 R^2} = \frac{r(1 - R^2)}{1 - rR}$$
$$\leq 2r(1 - R^2),$$

and

$$|z_1| > a - r\beta = \frac{R - r}{1 - rR} \geq c_2 > 0$$

for all r, $0 < r \leq \frac{1}{2}$ and $R \geq \frac{3}{4}$. Therefore, we also have $|z| \geq c_2$, and hence

$$1 - t = 1 - \frac{|z_1|}{|z|} \leq 1 - \frac{|z_1|^2}{|z|^2} = \frac{|z'|^2}{|z|^2} < (c')^2 r^2 (1 - R^2).$$

Also, since $|\theta| \leq c_3 |2 \sin \frac{\theta}{2}| = c_3 |1 - e^{i\theta}|$,

$$|\theta| \leq c_3 \left| 1 - \frac{z_1}{|z_1|} \right| \leq c_4 ||z_1| - z_1|$$

$$\leq c_4(||z_1| - R| + |R - |z_1||) \leq 2\, c_4 |z_1 - R| < c' r (1 - R^2)$$

for an appropriate choice of c'. Thus $\zeta \in N_{c'r}^R$. $\square$

As in [Ul2], for μ a nonnegative measure on B, $0 < r \leq \frac{1}{2}$, $\zeta \in S$, let

$$(8.13) \qquad\qquad M_r \mu(\zeta) = \sup_{0 < \rho < 1} \mu(E_r(\rho\zeta)).$$

The following Lemma is a key step in the proof of Proposition 8.3:

Lemma 8.7. *Let μ be a nonnegative regular Borel measure on B satisfying (8.2). Then there exists a constant C, independent of r and μ, such that for $0 < r \leq \frac{1}{2}$,*

$$\sigma(\{\zeta \in S : M_r \mu(\zeta) > \lambda\}) \leq C \frac{r^{2n-1}}{\lambda} \int_B (1 - |w|^2)^n \, d\mu(w)$$

for all $\lambda > 0$.

Proof. Define the finite measure $\mu^\star$ on B by

$$(8.14) \qquad\qquad d\mu^\star(w) = (1 - |w|^2)^n \, d\mu(w)$$

Fix $\lambda > 0$ and let

$$E = \{z \in B : \mu(E_r(z)) > \lambda\}.$$

Since $r \leq \frac{1}{2}$, by Lemma 8.5(c) $(1 - |w|^2) \geq \frac{1}{3}(1 - |z|^2)$ for all $w \in E_r(z)$. Therefore for all $z \in E$,

$$\mu^\star(E_r(z)) \geq c\,(1 - |z|^2)^n \lambda,$$

where c is a constant independent of r. As in Lemma 7.5, by Lemma 8.5(b), we can find a countable collection of points $\{z_\alpha\} \subset E$ such that the collection $\{E_r(z_\alpha)\}$ is pairwise disjoint, and

$$E \subset \bigcup_\alpha E_{3r}(z_\alpha).$$

Although Lemma 7.5 was proved for finite collections, it's extension to countable collections is clear.

Set $\zeta_\alpha = z_\alpha/|z_\alpha|$ and $\rho_\alpha = |z_\alpha|$. Also, let $V_\alpha = V_{3r}^{\rho_\alpha}(\zeta_\alpha)$. Suppose $M_r\mu(\zeta) > \lambda$. Then there exists a ρ such that $\rho\zeta \in E$. But then $\rho\zeta \in E_{3r}(z_\alpha)$ for some α. Therefore,

$$\{\zeta \in S : M_r\mu(\zeta) > \lambda\} \subset \bigcup_\alpha V_\alpha,$$

and as a consequence of Lemma 8.6,

$$\sigma(\{\zeta \in S : M_r\mu(\zeta) > \lambda\}) \le \sum_\alpha \sigma(V_\alpha) \le C\, r^{2n-1} \sum_\alpha (1 - \rho_\alpha^2)^n$$

$$\le C\frac{r^{2n-1}}{\lambda} \sum \mu^*(E_r(z_\alpha))$$

$$\le C\frac{r^{2n-1}}{\lambda} \int_B (1 - |w|^2)^n \, d\mu(w),$$

which proves the result. The last inequality in the above follows since the collection $\{E_r(z_\alpha)\}$ is pairwise disjoint. $\square$

Definition. For $\zeta \in S$, F a function defined on B, define the **radial maximal function** of F by

(8.15) $$(M_{\text{rad}}F)(\zeta) = \sup_{0 < \rho < 1} |F(\rho\zeta)|.$$

Lemma 8.8. Let μ be a nonnegative regular Borel measure on B satisfying (8.2), and let V_μ be defined by (8.3). Then there exists a positive constant C independent of μ such that

$$\sigma(\{\zeta \in S : (M_{\text{rad}}V_\mu)(\zeta) > \lambda\}) \le \frac{C}{\lambda} \int_B (1 - |w|^2)^n \, d\mu(w)$$

for all $\lambda > 0$.

Proof. Assume first that $n > 1$. Then by Lemma 6.6, there exists a constant C_1 such that

$$V_\mu(\rho\zeta) = \int_{E(\rho\zeta)} G(\rho\zeta, w) \, d\mu(w) \le C_1 \int_{E(\rho\zeta)} |\varphi_w(\rho\zeta)|^{-2(n-1)} \, d\mu(w).$$

Let $r_j = 2^{-j/(2n-2)}$. Then

$$E(\rho\zeta) = \bigcup_{j=(2n-2)}^\infty \left(E_{r_j}(\rho\zeta) \sim E_{r_{j+1}}(\rho\zeta) \right).$$

If $w \in E_{r_j}(\rho\zeta) \sim E_{r_{j+1}}(\rho\zeta)$, then

$$|\varphi_w(\rho\zeta)|^{2n-2} \geq (r_{j+1})^{2n-2} = 2^{-(j+1)}.$$

Therefore for all ρ, $0 < \rho < 1$,

$$V_\mu(\rho\zeta) \leq 2C_1 \sum_{j=2n-2}^{\infty} 2^j \mu(E_{r_j}(\rho\zeta))$$

$$\leq 2C_1 \sum_{j=2n-2}^{\infty} 2^j M_{r_j}\mu(\zeta),$$

and as a consequence,

$$(M_{\text{rad}}V_\mu)(\zeta) \leq 2C_1 \sum_{j=2n-2}^{\infty} 2^j M_{r_j}\mu(\zeta).$$

Choose α, $0 < \alpha < 1/(2n-2)$. Since $\sum 2^{-\alpha j} < \infty$,

$$(M_{\text{rad}}V_\mu)(\zeta) \leq 2C_1 \sum_{j=2n-2}^{\infty} 2^{-\alpha j} 2^{(1+\alpha)j} M_{r_j}\mu(\zeta)$$

$$\leq C_2 \sup_{j \geq (2n-2)} 2^{(1+\alpha)j} M_{r_j}\mu(\zeta).$$

For $\lambda > 0$ let

$$E = \{\zeta \in S : (M_{\text{rad}}V_\mu)(\zeta) > \lambda\}.$$

By the above,

$$E \subset \bigcup_{j=2n-2}^{\infty} \{\zeta \in S : M_{r_j}\mu(\zeta) > \frac{\lambda}{C_2} 2^{-(1+\alpha)j}\}.$$

Thus by Lemma 8.7, there exists a constant C_3 such that

$$\sigma(E) \leq \left(\frac{C_3}{\lambda} \int_B (1 - |w|^2)^n \, d\mu(w)\right) \left(\sum_{j=2n-2}^{\infty} 2^{(1+\alpha)j} r_j^{(2n-1)}\right)$$

$$= \left(\frac{C_3}{\lambda} \int_B (1 - |w|^2)^n \, d\mu(w)\right) \left(\sum_{j=2n-2}^{\infty} 2^{-\beta j}\right),$$

where

$$\beta = \frac{2n-1}{2n-2} - (1+\alpha) = \frac{1 - \alpha(2n-2)}{2n-2}.$$

By the above choice of α, $\beta > 0$. Therefore $\sum 2^{-\beta j} < \infty$, and the result now follows.

Suppose $n = 1$. In this case set $r_j = 2^{-j}$. By Lemma 6.6 there exists a constant C_1 such that

$$V_\mu(\rho\zeta) \leq C_1 \int_{E(\rho\zeta)} \log \frac{1}{|\varphi_w(\rho\zeta)|}\, d\mu(w).$$

As above,

$$V_\mu(\rho\zeta) \leq C_1 \log 2 \sum_{j=1}^{\infty} (j+1)\mu(E_{r_j}(\rho\zeta)).$$

Let $0 < \alpha < 1$. Since $\sum (j+1)2^{-\alpha j} < \infty$, we obtain as above that

$$(M_{\mathrm{rad}} V_\mu)(\zeta) \leq C_2 \sup_{j\geq 1} 2^{\alpha j} M_{r_j}\mu(\zeta).$$

Therefore by Lemma 8.7,

$$\sigma(E) \leq \frac{C_3}{\lambda} \left(\int_B (1 - |w|^2)\, d\mu(w) \right) \left(\sum_{j=1}^{\infty} r_j 2^{\alpha j} \right)$$

$$= \frac{C}{\lambda} \int_B (1 - |w|^2)\, d\mu(w).$$

In the above, the last step follows since

$$\sum_{j=1}^{\infty} r_j 2^{\alpha j} = \sum_{j=1}^{\infty} 2^{-j(1-\alpha)} < \infty. \qquad \square$$

Proof of Proposition 8.3. Let μ be a nonnegative regular Borel measure on B satisfying (8.2), and let V_1 be defined by (8.3). Let $\epsilon > 0$ be arbitrary, and let C be the constant of Lemma 8.8. Choose R, $0 < R < 1$, such that

$$C \int_{A_R} (1 - |w|^2)^n\, d\mu(w) < \epsilon^2.$$

Let μ_R denote the measure μ restricted to A_R, and let

$$V_R(z) = \int_{E(z)} G(z, w)\, d\mu_R(w).$$

By Lemma 8.5(d), there exists R' such that $E(z) \subset A_R$ for all z, $|z| \geq R'$. Thus $V_1(z) = V_R(z)$ for all z, $|z| \geq R'$. Therefore, by Lemma 8.8,

$$\sigma(\{\zeta \in S : \limsup_{r\to 1} V_1(r\zeta) > \epsilon\}) \leq \sigma(\{\zeta \in S : (M_{\mathrm{rad}} V_R)(\zeta) > \epsilon\})$$

$$\leq \frac{C}{\epsilon} \int_{A_R} (1 - |w|^2)^n\, d\mu(w) < \epsilon.$$

Since $\epsilon > 0$ was arbitrary, the result follows.   □

Example: We now give an example of a measure μ satisfying (8.2) for which

$$\limsup_{\substack{z \to \zeta \\ z \in D_\alpha(z)}} G_\mu(z) = +\infty$$

for every $\zeta \in S$ and $\alpha > 1$. Let $\{z_j\}$ be a countable subset of B with $|z_j| \to 1$ such that $D_\alpha(\zeta)$ contains infinitely many z_j for every $\zeta \in S$ and $\alpha > 1$. Such a sequence can be obtained by taking a countable dense subset $\{\zeta_j\}$ of S and a sequence $\{r_j\}$ increasing to 1, and setting $z_j = r_j \zeta_j$. Choose $c_j > 0$ such that

$$\sum_{j=1}^{\infty} c_j (1 - |z_j|^2)^n < \infty.$$

Define the measure μ on B by

$$\mu = \sum_{j=1}^{\infty} c_j \delta_{z_j},$$

where δ_a denotes pointmass measure at a. Then G_μ is a potential on B satisfying

$$G_\mu(z_j) = \infty \qquad \text{for all} \quad j.$$

8.2. Admissible Limits of Potentials.

In this section we consider sufficient conditions for the existence of admissible limits of potentials and $\mathcal{M}$-subharmonic functions. When $n = 1$, sufficient conditions for the existence of non-tangential limits of subharmonic functions in the unit disc were first provided by M. Arsove and A. Huber [AH]. The results were subsequently extended to the ball in $\mathbb{C}^n$ by J. Cima and C. S. Stanton [CiS].

If f is a nonnegative measurable function on B satisfying

(8.16)
$$\int_B (1 - |w|^2)^n f(w) \, d\lambda(w) < \infty,$$

we set

(8.17)
$$G_f(z) = \int_B G(z, w) f(w) \, d\lambda(w).$$

The $\mathcal{M}$-superharmonic function G_f is called the **Green potential** of f. The theorem we will prove is as follows:

Theorem 8.9. *([CiS]) Let f be a nonnegative measurable function on B satisfying (8.16). If in addition*

$$(8.18) \qquad \int_B (1 - |w|^2)^n f^p(w) \, d\lambda(w) < \infty$$

for some $p > n$, then for every $\alpha > 1$,

$$\lim_{\substack{z \to \zeta \\ z \in D_\alpha(\zeta)}} G_f(z) = 0 \qquad \text{for a.e.} \quad \zeta \in S.$$

As a consequence of the Riesz decomposition theorem and Theorems 7.1 and 8.9 we also have the following:

Corollary 8.10. *Let f be $\mathcal{M}$-subharmonic on B satisfying (8.1). If the Riesz measure μ_f of f is absolutely continuous and satisfies*

$$\int_B (1 - |w|^2)^n (\tilde{\Delta} f(w))^p \, d\lambda(w) < \infty$$

for some $p > n$, then f has admissible limits at a.e. $\zeta \in S$.

For the proof of Theorem 8.9 we need the following preparatory lemmas. As before, let $E(z) = \varphi_z(B_{\frac{1}{2}})$.

Lemma 8.11. *Let $\alpha > 1$, $\zeta \in S$. Then for all $z \in D_\alpha(\zeta)$,*

$$E(z) \subset D_\beta(\zeta) \qquad \text{for all} \quad \beta \geq 3\alpha.$$

Proof. Suppose $z \in B$, $\zeta \in S$. By identity (1.15)

$$1 - \langle \varphi_z(w), \zeta \rangle = 1 - \langle \varphi_z(w), \varphi_z(\varphi_z(\zeta)) \rangle$$
$$= \frac{(1 - |z|^2)(1 - \langle w, \varphi_z(\zeta) \rangle)}{(1 - \langle w, z \rangle)(1 - \langle z, \varphi_z(\zeta) \rangle)}.$$

But by identity (1.15) again

$$1 - \langle z, \varphi_z(\zeta) \rangle = 1 - \langle \varphi_z(0), \varphi_z(\zeta) \rangle = \frac{(1 - |z|^2)}{1 - \langle z, \zeta \rangle}.$$

Therefore

$$1 - \langle \varphi_z(w), \zeta \rangle = \frac{(1 - \langle z, \zeta \rangle)(1 - \langle w, \varphi_z(\zeta) \rangle)}{1 - \langle w, z \rangle}.$$

Suppose $w \in B_{\frac{1}{2}}$ and $z \in D_\alpha(\zeta)$. Then by (1.16) and the above,

$$
\begin{aligned}
|1 - \langle \varphi_z(w), \zeta \rangle| &\leq \frac{\alpha}{2}(1 - |z|^2)\frac{|1 - \langle w, \varphi_z(\zeta) \rangle|}{|1 - \langle w, z \rangle|} \\
&= \frac{\alpha}{2}(1 - |\varphi_z(w)|^2)\frac{|1 - \langle w, z \rangle||1 - \langle w, \varphi_z(\zeta) \rangle|}{(1 - |w|^2)} \\
&\leq \frac{\alpha}{2}\left(\frac{1 + |w|}{1 - |w|}\right)(1 - |\varphi_z(w)|^2) \\
&\leq \frac{3\alpha}{2}(1 - |\varphi_z(w)|^2),
\end{aligned}
$$

which proves the result. $\square$

For $z \in B$, $\alpha > 1$, define

(8.19) $$\tilde{D}_\alpha(z) = \{\zeta \in S : z \in D_\alpha(\zeta)\}.$$

Lemma 8.12. *Let $z \in B$, $z \neq 0$ and $\alpha > 1$. Set $\eta = z/|z|$. Then*

(a) $\tilde{D}_\alpha(z) \subset Q(\eta, 2\alpha(1 - |z|^2))$ and $\sigma(\tilde{D}_\alpha(z)) \leq C(1 - |z|^2)^n$.

(b) *There exists $c > 0$ and $R = R(\alpha, c)$, $0 < R < 1$, such that*

$$Q(\eta, c(1 - |z|^2)) \subset \tilde{D}_\alpha(z) \quad \text{and} \quad \sigma(\tilde{D}_\alpha(z)) \geq C(1 - |z|^2)^n$$

for all z, $|z| > R$.

Proof. Let $\eta = z/|z|$. If $\zeta \in \tilde{D}_\alpha(z)$, then

$$
\begin{aligned}
|1 - \langle \zeta, \eta \rangle|^{\frac{1}{2}} &\leq |1 - \langle z, \zeta \rangle|^{\frac{1}{2}} + |1 - \langle z, \eta \rangle|^{\frac{1}{2}} \\
&\leq |1 - \langle z, \zeta \rangle|^{\frac{1}{2}} + |1 - |z||^{\frac{1}{2}} \leq 2|1 - \langle z, \zeta \rangle|^{\frac{1}{2}}.
\end{aligned}
$$

Therefore,

$$|1 - \langle \zeta, \eta \rangle| < 2\alpha(1 - |z|^2),$$

which proves the first part of (a). The second part follows by Lemma 7.2.

To prove (b), suppose $\zeta \in Q(\eta, c(1-|z|^2))$, where $c > 0$ is to be determined. Then

$$
\begin{aligned}
|1 - \langle z, \zeta \rangle| &\leq |1 - \langle \zeta, \eta \rangle| + |\langle \zeta, \eta - z \rangle| \\
&\leq c(1 - |z|^2) + (1 - |z|) \\
&\leq \left(c + \frac{1}{1 + |z|}\right)(1 - |z|^2).
\end{aligned}
$$

Since $\alpha > 1$, we can choose $c > 0$ and $0 < R < 1$ such that

$$\left(c + \frac{1}{1 + |z|}\right) < \frac{\alpha}{2} \qquad \text{for all} \quad z, |z| > R,$$

which proves (b). □

For a nonnegative measure μ on B, we consider the following variation of the square area integral (7.23): for $\alpha > 1$, set

$$(8.20) \qquad\qquad S_\alpha^* \mu(\zeta) = \mu(D_\alpha(\zeta)).$$

Proposition 8.13. *Let μ be a nonnegative regular Borel measure on B. Then*

$$\int_B (1 - |w|^2)^n \, d\mu(w) < \infty \qquad \text{if and only if} \qquad S_\alpha^* \mu \in L^1(S)$$

for all $\alpha > 1$.

Proof. For a set E, let χ_E denote the characteristic function of E. Then by Tonelli's theorem,

$$\int_S S_\alpha^* \mu(\zeta) \, d\sigma(\zeta) = \int_S \int_B \chi_{D_\alpha(\zeta)}(z) \, d\mu(z) \, d\sigma(\zeta)$$

$$= \int_B \int_S \chi_{\tilde{D}_\alpha(z)}(\zeta) \, d\sigma(\zeta) \, d\mu(z)$$

$$= \int_B \sigma(\tilde{D}_\alpha(z)) \, d\mu(z).$$

By Lemma 8.12, $\sigma(\tilde{D}_\alpha(z)) \le C(1 - |z|^2)^n$, and thus $S_\alpha^* \mu \in L^1(S)$ if μ satisfies (8.2).

Conversely, since μ is regular, μ satisfies (8.2) if and only if $\int_{A_R}(1 - |z|^2)^n \, d\mu(z) < \infty$ for some R, $0 < R < 1$. Thus by Lemma 8.12 (b) and the above,

$$\int_{A_R} (1 - |z|^2)^n \, d\mu(z) \le C \int_{A_R} \sigma(\tilde{D}_\alpha(z)) \, d\mu(z)$$

$$\le C \int_S S_\alpha^* \mu(\zeta) \, d\sigma(\zeta). \qquad \square$$

Corollary 8.14. *Let μ be a nonnegative Borel measure on B satisfying (8.2). Then for every $\alpha > 1$,*

$$\mu(D_\alpha(\zeta)) < \infty \qquad \text{for a.e. } \zeta \in S,$$

and thus

$$\lim_{r \to 1} \mu(D_\alpha(\zeta) \cap A_r) = 0 \qquad \text{for a.e. } \zeta \in S.$$

Proof. The result is an immediate consequence of the fact that the function $S_\alpha^* \mu \in L^1(S)$. □

Proof of Theorem 8.9. Suppose f satisfies (8.16) and (8.18). Write $G_f(z) = V_1(z) + V_2(z)$, where V_1 and V_2 are defined by (8.3) and (8.4) respectively. Since V_2 has admissible limit 0 a.e., we only need to prove the result for V_1.

We first note that for $n > 1$, by Lemma 6.6 and integration in polar coordinates,

$$\int_{B_{\frac{1}{2}}} g(z)^q \, d\lambda(z) \le C_1 \int_{B_{\frac{1}{2}}} |z|^{-q(2n-2)} \, d\lambda(z)$$

$$\le C_2 \int_0^{\frac{1}{2}} r^{2n-q(2n-2)-1} \, dr < \infty$$

provided $2n - q(2n-2) > 0$, i.e., $q < n/(n-1)$. If $n = 1$, the above holds for all $q < \infty$. Thus

$$(8.21) \qquad \sup_{z \in B} \int_{E(z)} G^q(z,w) \, d\lambda(w) < \infty \qquad \text{for all} \quad q < \frac{n}{n-1}.$$

Suppose f satisfies (8.18) for some $p > n$. Let $q = p/(p-1)$ be the conjugate exponent of p. Since $p > n$, $q < n/(n-1)$. Thus by Hölder's inequality and the above,

$$V_1(z) = \int_{E(z)} G(z,w) f(w) \, d\lambda(w) \le C \left[\int_{E(z)} f^p(w) \, d\lambda(w) \right]^{1/p}.$$

Suppose $z \in D_\alpha(\zeta)$, $\alpha > 1$. Then by Lemma 8.5(d) and Lemma 8.11, $E(z) \subset D_\beta(z) \cap A_r$ for any $\beta \ge 3\alpha$ and $r^2 = 3|z|^2 - 2$. Thus

$$V_1(z) \le C \left[\int_{D_\beta(\zeta) \cap A_r} f^p(w) \, d\lambda(w) \right]^{1/p}.$$

Therefore, by Corollary 8.14 with $d\mu = f^p \, d\lambda$, the function V_1 has admissible limit 0 at a.e. $\zeta \in S$. □

Example: We now provide an example to show that the exponent $p > n$ in the hypothesis of Theorem 8.9 is best possible. Specifically, we construct a nonnegative measurable function f satisfying (8.16) and (8.18) with $p = n$ for which

$$(8.22) \qquad \limsup_{\substack{z \to \zeta \\ z \in D_\alpha(\zeta)}} G_f(z) = +\infty \qquad \text{for every} \quad \zeta \in S.$$

As in the example of the previous section, let $\{z_j\}$ be a countable subset of B with $|z_j| \to 1$ such that $D_\alpha(z)$ contains infinitely many z_j for every $\zeta \in S$ and $\alpha > 1$. For each j, choose r_j, $0 < r_j < \frac{1}{2}$ and $c_j > 0$ such that

(a)   the family $\{E(z_j, r_j)\}$ is pairwise disjoint, and

(b) $\sum_{j=1}^{\infty} (1 - |z_j|^2)(c_j)^{1/n} < \infty$.

Let $\{a_j\}$ be a sequence of positive numbers with $a_j \to \infty$. For each j, choose a nonnegative measurable function f_j satisfying

(1)   support of $f_j \subset E(z_j, r_j)$,

(2) $\int f_j^n \, d\lambda < c_j$, and

(3) $\int_B f_j(w) G(z_j, w) \, d\lambda(w) > a_j$.

It is clear that for each j we can find f_j satisfying (1) and (2). Fix a j, and set $E_j = E(z_j, r_j)$. If we cannot find an f_j satisfying (1) - (3), then we have

$$\int_{E_j} f(w) G(z_j, w) \, d\lambda(w) \le a_j \, (c_j)^{1/n}$$

for all nonnegative measurable functions f satisfying $\int_{E_j} f^n \, d\lambda \le 1$. But this implies that $G(z_j, w) \in L^{n/(n-1)}(E_j, \lambda)$, which is a contradiction. Thus there exists a nonnegative measurable function satisfying (1) - (3). In fact, one can choose f_j to be C_c^{∞}.

Let $f = \sum f_j$. We now show that f satisfies (8.16) and (8.18). First, by (4.4),

$$\lambda(E(z_j, r_j)) = \frac{r_j^{2n}}{(1 - r_j^2)^n} \le C_n.$$

Since $(1 - |w|^2) \le 3(1 - |z_j|^2)$ for $w \in E(z_j, r_j)$, by Hölder's inequality,

$$\int_B (1 - |w|^2)^n f(w) \, d\lambda(w) \le 3 \sum_{j=1}^{\infty} (1 - |z_j|^2)^n \int_{E(z_j, r_j)} f_j \, d\lambda$$

$$\le C \sum_{j=1}^{\infty} (1 - |z|_j^2)^n (c_j)^{\frac{1}{n}}$$

$$\le C \sum_{j=1}^{\infty} (1 - |z_j|^2)(c_j)^{\frac{1}{n}} < \infty.$$

Also,

$$\int_B (1 - |w|^2)^n f^n(w) \, d\lambda(w) \le 3 \sum_{j=1}^{\infty} (1 - |z_j|^2)^n c_j.$$

Since $l^1 \subset l^n$, the above series also converges. Thus f satisfies both (8.16) and (8.18) with $p = n$. Finally, since

$$G_f(z_j) \geq \int_B f_j(w) G(z_j, w) \, d\lambda(w) \geq a_j,$$

G_f satisfies (8.22) at every $\zeta \in S$. By choosing the f_j to be C_c^∞, we also obtain the existence of a C^∞ M-superharmonic function satisfying (8.22).

Our final theorem of this section, although valid for all $n \geq 1$, is of particular interest when $n = 1$.

Proposition 8.15. Let $\{a_j\}$ be a sequence in B satisfying

$$(8.23) \qquad \sum_{j=1}^\infty (1 - |a_j|^2)^n < \infty,$$

and let $\mu = \sum_{j=1}^\infty \delta_{a_j}$, where δ_a denotes pointmass measure at a. Then

$$\lim_{\substack{z \to \zeta \\ z \in D_\alpha(\zeta)}} G_\mu(z) = 0 \qquad \text{for a.e. } \zeta \in S.$$

Proof. As a consequence of Proposition 8.2, we again only need to consider the function $V_1(z)$. For each $\alpha > 1$,

$$\mu(D_\alpha(\zeta)) = |\{j \in N : a_j \in D_\alpha(\zeta)\}|,$$

where for J a subset of N, $|J|$ denotes the number of elements in J. If $\mu(D_\alpha(\zeta)) < \infty$, then $D_\alpha(\zeta)$ contains only a finite number of a_j. Consequently for such a ζ,

$$\lim_{\substack{z \to \zeta \\ z \in D_\alpha(\zeta)}} V_1(z) = 0.$$

But by Corollary 8.14, $\mu(D_\alpha(\zeta)) < \infty$ for a.e. $\zeta \in S$. $\square$

When $n = 1$, and the sequence $\{a_j\}$ satisfies (8.23), then

$$G_\mu(z) = \sum_{j=1}^\infty \log \left| \frac{1 - \bar{a}_j z}{z - a_j} \right| = -\log |B(z)|,$$

where B is the Blaschke product with zeros $\{a_j\}$ given by

$$B(z) = \prod_{j=1}^\infty \frac{|a_j|}{a_j} \frac{z - a_j}{1 - \bar{a}_j z}.$$

The hypothesis (8.23) when $n = 1$ is known as the Blaschke condition, which is necessarry and sufficient for the existence of a bounded analytic function in U having prescribed zeros $\{a_j\}$.

Remark: There is one more application of Proposition 8.13 which is worth mentioning. If f is $\mathcal{M}$-harmonic on B, then by Theorem 6.18, $f \in \mathcal{H}^2(B)$ if and only if

$$\int_B (1 - |z|^2)^n |\widetilde{\nabla} f(z)|^2 \, d\lambda(z) < \infty.$$

As in the proof of Theorem 6.18, the Riesz measure μ of $|f|^2$ is given by $d\mu(z) = 2|\widetilde{\nabla} f(z)|^2 \, d\lambda(z)$. Thus the area integral of f, as defined by (7.23) satisfies

$$S_\alpha^2 f(\zeta) = \int_{D_\alpha(\zeta)} |\widetilde{\nabla} f(z)|^2 \, d\lambda(z) = \frac{1}{2} S_\alpha^* \mu(\zeta).$$

As a consequence, $f \in \mathcal{H}^2(B)$ if and only if $S_\alpha f \in L^2(S)$.

8.3. Tangential Limits of Potentials.

In this section we explore sufficient conditions for the existence of tangential boundary limits.

Definition. For $c > 0$, $\tau > 1$ and $\zeta \in S$, set

(8.24) $$T_{\tau,c}(\zeta) = \{z \in B : |1 - \langle z, \zeta \rangle|^\tau < c(1 - |z|^2)\} \ .$$

When $\tau = 1$ and $c > \frac{1}{2}$ we obtain the admissible approach regions. For $\tau > 1$, the regions $T_{\tau,c}(\zeta)$ have tangential contact in all directions at the boundary point $\zeta \in S$. The number τ will be called the degree of tangency.

Definition. A function V on B is said to have T_τ-limit L at $\zeta \in S$ if

$$\lim_{\substack{z \to \zeta \\ z \in T_{\tau,c}(\zeta)}} V(z) = L$$

for every $c > 0$.

The main result is as follows:

Theorem 8.16. ([Sto7]) *If f is a nonnegative measurable function on B satisfying*

(8.25) $$\int_B (1 - |w|^2)^\alpha f^p(w) \, d\lambda(w) < \infty,$$

for some α, $0 < \alpha < n$, and some $p > n$, then G_f has T_τ-limit 0 at a.e. $\zeta \in S$ for all τ, $1 \leq \tau \leq \frac{n}{\alpha}$.

Notes: (1) If f satisfies (8.25) for some α, $0 < \alpha < n$ and some $p > 1$, then by Hölder's inequality,

$$\int_B (1 - |w|^2)^n f(w) \, d\lambda(w)$$
$$\leq \left[\int_B (1 - |w|^2)^{(n-\frac{\alpha}{p})q} \, d\lambda(w) \right]^{\frac{1}{q}} \left[\int_B (1 - |w|^2)^\alpha f^p(w) \, d\lambda(w) \right]^{\frac{1}{p}}.$$

Since $(n - \frac{\alpha}{p})q = (n - \alpha)q + \alpha > n$, the first integral above is finite. Thus f also satisifes (8.16) and hence G_f is a potential on B.

(2) If f satisfies (8.25) for the other extreme, namely $\alpha = 0$, then it is easy to show that
$$\lim_{z \to \zeta} G_f(z) = 0$$

for all $\zeta \in S$.

(3) The analogue of Theorem 8.16 for euclidean Green potentials in a half-space in $\mathbb{R}^n$ has been proved by Jang-Wei Wu [Wu].

Before proving the result, we need some preparatory lemmas. As in the previous section, for $z \in B$, $z \neq 0$, define the subset $\widetilde{T}_{\tau,c}(z)$ of S

$$\widetilde{T}_{\tau,c}(z) = \{\zeta \in S : z \in T_{\tau,c}(\zeta)\}.$$

The proof of the following lemma is similar to the proofs of the corresponding results in the previous section, and thus is ommitted.

Lemma 8.17.

(a) If $z \in T_{\tau,c}(\zeta)$, $\tau > 1$, then

$$E(z) \subset T_{\tau,c'}(\zeta) \qquad \text{for any} \quad c' \geq c \, 3^{\tau+1}.$$

(b) If $z \in T_{\tau,c}(\zeta)$, then

$$\sigma(\widetilde{T}_{\tau,c}(z)) \leq C \, (1 - |z|^2)^{n/\tau}.$$

(c) If μ satisfies

(8.26) $$\int_B (1 - |z|^2)^{n/\tau} \, d\mu(z) < \infty$$

for some $\tau \geq 1$, then for all $c > 0$, $\mu(T_{\tau,c}(\zeta)) < \infty$ for a.e. $\zeta \in S$, and

$$\lim_{r \to 1} \mu(T_{\tau,c}(\zeta) \cap A_r) = 0 \qquad \text{for a.e. } \zeta \in S.$$

For $z \in B$, c real, and $\alpha > n$ consider the following integrals:

$$I_c(z) = \int_S \frac{1}{|1 - \langle z, \zeta \rangle|^{n+c}} d\sigma(\zeta),$$

$$J_{c,\alpha}(z) = \int_B \frac{(1 - |w|^2)^\alpha}{|1 - \langle z, w \rangle|^{\alpha+c}} d\lambda(w).$$

Although the following asymptotic estimates are well known ([Ru3; Proposition 1.4.10]), we include the result for completeness.

Proposition 8.18. *For* $\alpha > n$, $z \in B$,

$$I_c(z) \approx J_{c,\alpha}(z) \approx \begin{cases} (1 - |z|^2)^{-c}, & c > 0, \\ \log \dfrac{1}{(1 - |z|^2)}, & c = 0, \\ 1 & , & c < 0. \end{cases}$$

Proof. We first prove the result for $I_c(z)$. If $c \leq -n$, then clearly $I_c(z)$ is bounded on B. Thus assume $c > -n$, and write $n + c = 2\beta$, $\beta > 0$. Since $\langle z, \zeta \rangle^k$ and $\langle z, \zeta \rangle^m$ are orthogonal in $L^2(S)$, by the binomial expansion

$$I_c(z) = \int_S \frac{d\sigma(\zeta)}{|1 - \langle z, \zeta \rangle|^{2\beta}}$$

$$= \sum_{k=0}^{\infty} \frac{\Gamma^2(k + \beta)}{\Gamma^2(\beta)(k!)^2} \int_S |\langle z, \zeta \rangle|^{2k} d\sigma(\zeta).$$

By Lemma 2.4 and unitary invariance,

$$\int_S |\langle z, \zeta \rangle|^{2k} d\sigma(\zeta) = |z|^{2k} \int_S |\zeta_1^k|^2 d\sigma(\zeta) = \frac{\Gamma(n) k!}{\Gamma(n + k)} |z|^{2k}.$$

Therefore,

(8.27) $$I_c(z) = \frac{\Gamma(n)}{\Gamma^2(\beta)} \sum_{k=0}^{\infty} \frac{\Gamma^2(k + \beta)}{k! \Gamma(n + k)} |z|^{2k}.$$

Since

$$\lim_{k \to \infty} k^{b-a} \frac{\Gamma(k + a)}{\Gamma(k + b)} = 1,$$

we have

$$\frac{\Gamma^2(k+\beta)}{k!\,\Gamma(k+n)} \approx \frac{\Gamma(k+c)}{k!} \approx \frac{1}{k^{1-c}},$$

from which the result follows for $I_c(z)$.

For $J_{c,\alpha}(z)$, by integration in polar coordinates,

$$J_{c,\alpha}(z) = 2n \int_0^1 r^{2n-1}(1-r^2)^{\alpha-n-1} I_{\alpha-n+c}(rz)\,dr$$

which by (8.27)

$$= \frac{\Gamma(n+1)}{\Gamma^2(\beta)} \sum_{k=0}^\infty \frac{\Gamma^2(k+\beta)}{k!\,\Gamma(n+k)} |z|^{2k} 2 \int_0^1 r^{2n+2k-1}(1-r^2)^{\alpha-n-1}\,dr$$

$$= \frac{\Gamma(n+1)\Gamma(\alpha-n)}{\Gamma^2(\beta)} \sum_{k=0}^\infty \frac{\Gamma^2(k+\beta)}{k!\,\Gamma(k+\alpha)} |z|^{2k}$$

where $2\beta = \alpha + c$, provided $\alpha + c > 0$. If $\alpha + c \le 0$, then $J_{c,\alpha}(z)$ is obviously bounded on B. As above,

$$\frac{\Gamma^2(k+\beta)}{k!\,\Gamma(k+\alpha)} \approx \frac{\Gamma(k+c)}{k!} \approx \frac{1}{k^{1-c}},$$

from which the result now follows. $\square$

For $0 < \alpha < n$, $\zeta \in S$, and $z \in B$, define the functions k_ζ^α and K_z^α on S by

$$k_\zeta^\alpha(\eta) = \frac{1}{|1 - \langle \zeta, \eta \rangle|^\alpha} \quad \text{and} \quad K_z^\alpha(\eta) = \frac{1}{|1 - \langle z, \eta \rangle|^\alpha}.$$

Since $k_\zeta^\alpha(\eta) = \lim_{r \to 1} K_{r\zeta}^\alpha(\eta)$, by Fatou's lemma and Proposition 8.18, $k_\zeta^\alpha \in L^1(S)$ for all α, $0 < \alpha < n$. In fact, by [Ru3, p.54],

$$\|k^\alpha\|_1 = \|k_\zeta^\alpha\|_1 = \int_S |1 - t_1|^{-\alpha}\,d\sigma(t) = \frac{\Gamma(n)\Gamma(n-\alpha)}{\Gamma^2(n-\frac{\alpha}{2})}.$$

If μ is a finite measure on S, set

$$K^\alpha[\mu](z) = \int_S |1 - \langle z, \eta \rangle|^{-\alpha}\,d\mu(\eta).$$

Lemma 8.19. *Let $0 < \alpha < n$ and let μ be a finite measure on S. Then for every $\zeta \in S$,*

$$\int_S \frac{1}{|1 - \langle \eta, \zeta \rangle|^\alpha} \, d\mu(\eta) \le C_\alpha \, \|k^\alpha\|_1^{n-\alpha} \, M[\mu](\zeta).$$

where $M[\mu]$ is the maximal function of μ as defined in (7.6).

Proof. Fix $\zeta \in S$ and let $\delta = 2^N \|k^\alpha\|_1$, where N is the smallest integer such that $\delta > 2$. For $k = 0, 1, 2, ...$, let

$$V_k = \{t \in S : \frac{\delta}{2^{k+1}} < |1 - \langle t, \zeta \rangle| \le \frac{\delta}{2^k}\},$$

and

$$Q_k = \{t \in S : |1 - \langle t, \zeta \rangle| \le \frac{\delta}{2^k}\}.$$

Since $\mu(Q_k) \le \dfrac{\delta^n}{2^{nk}} M[\mu](\zeta)$, for $k = 0, 1, 2, ...,$

$$\int_{V_k} k_\zeta^\alpha(t) \, d\mu(t) \le \frac{2^{(k+1)\alpha}}{\delta^\alpha} \mu(Q_k) \le \frac{2^\alpha \delta^{n-\alpha}}{2^{(n-\alpha)k}} M[\mu](\zeta).$$

Summing over k gives the desired result. $\square$

The following Lemma, which is similar to Theorem 2.3 of [NRS], provides one of the key steps in proving the result.

Lemma 8.20. *Let $0 < \alpha < n$, $\zeta \in S$, and let μ be a finite measure on S. Then for all $z \in B$,*

$$K^\alpha[\mu](z) \le C \left[\frac{|1 - \langle z, \zeta \rangle|^n}{(1 - |z|^2)^\alpha} + \|k^\alpha\|_1^{n-\alpha} \right] M[\mu](\zeta).$$

Proof. Fix $\zeta \in S$ and $z \in B$. Set $\delta = |1 - \langle z, \zeta \rangle|$ and let $Q_{2\delta} = Q(\zeta, 2\delta)$, where Q is as defined by (5.5). Then

$$\int_{Q_{2\delta}} K_z^\alpha(\eta) \, d\mu(\eta) \le \frac{2^\alpha}{(1 - |z|^2)^\alpha} \mu(Q_{2\delta}) \le C \frac{\delta^n}{(1 - |z|^2)^\alpha} M[\mu](\zeta).$$

Suppose $\eta \in S \sim Q_{2\delta}$. Since

$$|1 - \langle \eta, \zeta \rangle|^{\frac{1}{2}} \le |1 - \langle z, \eta \rangle|^{\frac{1}{2}} + |1 - \langle z, \zeta \rangle|^{\frac{1}{2}},$$

we obtain

$$|1 - \langle z, \eta \rangle|^{\frac{1}{2}} \ge |1 - \langle \eta, \zeta \rangle|^{\frac{1}{2}} - \tfrac{1}{\sqrt{2}}|1 - \langle \eta, \zeta \rangle|^{\frac{1}{2}} \ge c |1 - \langle \eta, \zeta \rangle|^{\frac{1}{2}}$$

where $c > 0$. Therefore

$$\int_{S \sim Q_{2\delta}} K_z^\alpha(\eta) \, d\mu(\eta) \leq C \int_S k_\zeta^\alpha(\eta) \, d\mu(\eta).$$

The result now follows by Lemma 8.19. $\square$

Proof of Theorem 8.16. Suppose f is a nonnegative measurable function on B satisfying (8.25) for some $p > n$. Write $G_f(z) = V_1(z) + V_2(z)$, where V_1 and V_2 are defined as in (8.3) and (8.4) respectively. Since $T_{\tau',c}(\zeta) \subset T_{\tau,c'}(\zeta)$ for all $\tau' \leq \tau$ with $c' = c^{\tau/\tau'}$, it suffices to prove the results for $\tau = n/\alpha$.

We first consider the function V_1. As in the proof of Theorem 8.9, by Hölder's inequality,

$$V_1(z) \leq C \left[\int_{E(z)} f^p(w) \, d\lambda(w) \right]^{1/p}.$$

Suppose $z \in T_{\tau,c}(\zeta)$ with $\tau = n/\alpha$. Then by Lemma 8.17, $E(z) \subset T_{\tau,c'}(\zeta) \cap A_r$ for any $c' \geq c3^{\tau+1}$ and $r^2 = 3|z|^2 - 2$. Thus

$$V_1(z) \leq C \left[\int_{T_{\tau,c'}(\zeta) \cap A_r} f^p(w) \, d\lambda(w) \right]^{1/p}.$$

Hence by Lemma 8.17, V_1 has T_τ-limit 0 at a.e. $\zeta \in S$.

Consider the function $V_2(z)$. Let $0 < R < 1$. As in (8.6),

$$V_2(z) \leq C_R (1 - |z|^2)^n + C (1 - |z|^2)^n \int_{A_R} \frac{(1 - |w|^2)^n}{|1 - \langle z, w \rangle|^{2n}} f(w) \, d\lambda(w).$$

Suppose that f satisfies (8.25) for some $p > 1$. Then by Hölder's inequality

$$\int_{A_R} \frac{(1 - |w|^2)^n}{|1 - \langle z, w \rangle|^{2n}} f(w) \, d\lambda(w)$$

$$\leq \left[\int_{A_R} \frac{(1 - |w|^2)^{(n - \frac{\alpha}{p})q}}{|1 - \langle z, w \rangle|^{nq + (n - \frac{\alpha}{p})q}} d\lambda(w) \right]^{\frac{1}{q}} \left[\int_{A_R} \frac{(1 - |w|^2)^\alpha f^p(w)}{|1 - \langle z, w \rangle|^\alpha} d\lambda(w) \right]^{\frac{1}{p}},$$

which by Proposition 8.18

$$\leq C (1 - |z|^2)^{-n} \left[\int_{A_R} \frac{(1 - |w|^2)^\alpha f^p(w)}{|1 - \langle z, w \rangle|^\alpha} d\lambda(w) \right]^{\frac{1}{p}}.$$

Thus

$$V_2(z) \leq C_R(1 - |z|^2)^n + C \left[\int_{A_R} \frac{(1 - |w|^2)^\alpha f^p(w)}{|1 - \langle z, w \rangle|^\alpha} \, d\lambda(w) \right]^{\frac{1}{p}}.$$

As in (8.7), define the measure μ_R on S by

$$\int_S h(t) \, d\mu_R(t) = \int_{A_R} h(\tfrac{w}{|w|})(1 - |w|^2)^\alpha f^p(w) \, d\lambda(w).$$

Therefore, since $|1 - \langle z, w \rangle| \geq \frac{1}{2}|1 - \langle z, \tfrac{w}{|w|} \rangle|$,

(8.28) $$V_2(z) \leq C_R(1 - |z|^2)^n + (K^\alpha[\mu_R](z))^{\frac{1}{p}}.$$

If $z \in T_{\tau,c}(\zeta)$ with $\tau = n/\alpha$, then $|1 - \langle z, \zeta \rangle|^n \leq c^\alpha(1 - |z|^2)^\alpha$. Thus by Lemma 8.20,

$$K^\alpha[\mu_R](z) \leq C_\alpha M[\mu_R](\zeta), \qquad \text{for all} \quad z \in T_{\tau,c}(\zeta).$$

The result for V_2 now follows as in the proof of Proposition 8.2. $\square$

8.4. Convergence in L^p.

Our final convergence result involves convergence of potentials in L^p. This concept was initially considered by Ziomek [Zi] with regard to nontangential limits in $\mathbb{R}^n$, and subsequently by Cima and Stanton [CiS] for admissible limits in L^p.

For $\tau \geq 1$ and $c > 0$ ($c > \frac{1}{2}$ when $\tau = 1$) let $T_{\tau,c}(\zeta)$ be the tangential approach region defined in (8.24), and for $0 < \rho < 1$ set

$$T_{\tau,c,\rho}(\zeta) = T_{\tau,c}(\zeta) \cap A_\rho.$$

Definition. *A function V on B is said to have T_τ-limit L in L^p at $\zeta \in S$, if*

$$\lim_{\rho \to 1} \frac{1}{\nu(T_{\tau,c,\rho}(\zeta))} \int_{T_{\tau,c,\rho}(\zeta)} |V(z) - L|^p \, d\nu(z) = 0.$$

When $\tau = 1$, V is said to have **admissible limit** L in L^p.

Theorem 8.21. *If μ is a nonnegative Borel measure on B satisfying*

(8.29) $$\int_B (1 - |w|^2)^\alpha \, d\mu(w) < \infty$$

for some α, $0 < \alpha \leq n$, then G_μ has T_τ-limit 0 in L^q at a.e. $\zeta \in S$ for all $q < \frac{n}{n-1}$ and all τ, $1 \leq \tau \leq \frac{n}{\alpha}$.

If (8.29) holds for $\alpha < n$, then the measure μ also satisfies (8.29) with $\alpha = n$, and thus G_μ is a potential on B. If the measure μ is absolutely continuous with $d\mu = f d\lambda$, where f satisfies (8.25) for some $p > 1$, then we obtain the following:

Theorem 8.22. *Suppose f is a nonnegative measurable function satisfying (8.25) for some $p > 1$ and some α, $0 < \alpha \leq n$ (and (8.16) if $\alpha = n$). Then for all q,*

$$(8.30) \qquad 0 < q < \begin{cases} \dfrac{pn}{n-p}, & \text{if } p < n, \\ \infty, & p \geq n, \end{cases}$$

G_f has $\mathcal{T}_\tau$-limit 0 in L^q at a.e. $\zeta \in S$, for all τ, $1 \leq \tau \leq \frac{n}{\alpha}$.

Remark: Theorem 8.21 for $\alpha = n$ is due to Cima and Stanton [CiS]. The case $0 < \alpha < n$ and also Theorem 8.22 are due to the author [Sto7].

Proofs. As previously, we write $G_\mu = V_1 + V_2$, where V_1 and V_2 are defined by (8.3) and (8.4) respectively. Also, we will prove the result for the case $0 < \alpha < n$. The case of admissible limits when $\alpha = n$ is similar.

We begin by considering the function $V_2(z)$. Suppose μ satisfies (8.29) for some α, $0 < \alpha < n$. Since

$$|1 - \langle z, w \rangle|^{2n-\alpha} \geq C (1 - |z|^2)^n (1 - |w|^2)^{n-\alpha},$$

as in (8.6), for $0 < R < 1$, we obtain

$$V_2(z) \leq C_R (1 - |z|^2) + C \int_{A_R} \frac{(1 - |w|^2)^\alpha}{|1 - \langle z, w \rangle|^\alpha} \, d\mu(w).$$

where C_R is a constant depending on R. Thus by Lemma 8.20, if $z \in \mathcal{T}_{\tau,c}(\zeta)$,

$$V_2(z) \leq C_R(1 - |z|^2)^n + C M[\mu_R](\zeta),$$

where μ_R is defined as in (8.7). Therefore, for $q > 0$,

$$\frac{1}{\nu(\mathcal{T}_{\tau,c,\rho}(\zeta))} \int_{\mathcal{T}_{\tau,c,\rho}(\zeta)} V_2(z)^q \, d\nu(z) \leq C_R (1 - \rho^2)^{nq} + C \left(M[\mu_R](\zeta) \right)^q.$$

Thus V_2 has $\mathcal{T}_\tau$-limit 0 in L^q for all $q > 0$. If f satisfies (8.25) for some $p > 1$ and some α, $0 < \alpha < n$, then as a consequence of (8.28), the same result holds for V_2 in this case also.

We now turn our attention to $V_1(z)$. As we will see, the restriction on q is determined by $V_1(z)$. Suppose that μ is a measure satisfying (8.29) for some α, $0 < \alpha < n$. If $z \in T_{\tau,c}(\zeta)$, then by Lemma 8.17,

$$E(z) \subset T_{\tau,c',r}(\zeta), \quad c' = c3^{\tau+1}, \ r^2 = 3\rho^2 - 2.$$

Also, we note that $w \in E(z)$ if and only if $z \in E(w)$. Thus by the continuous version of Minkowski's inequality, for $1 \le q < \frac{n}{n-1}$,

$$\left(\int_{T_{\tau,c,\rho}(\zeta)} V_1(z)^q \, d\lambda(z) \right)^{1/q} \le \int_{T_{\tau,c',r}(\zeta)} \left[\int_{E(w)} G(z,w)^q \, d\lambda(z) \right]^{1/q} d\mu(w)$$

$$\le C \, \mu(T_{\tau,c',r}(\zeta)).$$

Thus by Lemma 8.17, if $\tau = n/\alpha$,

$$\lim_{\rho \to 1} \int_{T_{\tau,c,\rho}(\zeta)} V_1(z)^q \, d\lambda(z) = 0$$

for a.e. $\zeta \in S$. Since $\nu(T_{\tau,c,\rho}(\zeta)) \approx (1 - \rho^2)^{\frac{n}{\tau}+1}$,

$$\frac{1}{\nu(T_{\tau,c,\rho}(\zeta))} \int_{T_{\tau,c,\rho}(\zeta)} V_1(z)^q \, d\nu(z) \le \int_{T_{\tau,c,\rho}(\zeta)} V_1(z)^q \, d\lambda(z).$$

Therefore, for all τ, $1 \le \tau \le n/\alpha$, V_1 has T_τ-limit 0 in L^q at a.e. $\zeta \in S$ for all q, $1 \le q < \frac{n}{n-1}$. The result for $0 < q < 1$ follows by Hölder's inequality.

Finally, suppose f satisfies (8.25) for some $p > 1$ and some α, $0 < \alpha < n$. Let $0 \le \beta \le 1$ be such that

$$(8.31) \qquad\qquad (n - p) < \beta \, p(n - 1)$$

Let $s = p/(p-1)$ be the conjugate exponent of p. If β satisfies (8.31), then $s(1 - \beta) < \frac{n}{n-1}$. Thus by Hölder's inequality

$$V_1(z) \le \left[\int_{E(z)} G(z,w)^{s(1-\beta)} \, d\lambda(w) \right]^{1/s} \left[\int_{E(z)} G(z,w)^{\beta p} f(w)^p \, d\lambda(w) \right]^{1/p}$$

$$\le C \left[\int_{E(z)} G(z,w)^{\beta p} f(w)^p \, d\lambda(w) \right]^{1/p}.$$

If $q \ge p$, then as above, by the continuous version of Minkowski's inequality,

$$\int_{T_{\tau,c,\rho}(\zeta)} V_1(z)^q \, d\lambda(z)$$

$$\le C \left[\int_{T_{\tau,c',r}(\zeta)} f(w)^p \left[\int_{E(w)} G(z,w)^{\beta q} d\lambda(z) \right]^{\frac{q}{q}} d\lambda(w) \right]^{\frac{q}{p}}.$$

If β is such that $\beta q < \frac{n}{n-1}$, then

$$\int_{E(w)} G(z,w)^{\beta q} d\lambda(z) \leq C.$$

If $p < n$ and $p \leq q < np/(n-p)$, then we can choose a $\beta < 1$ such that

$$\frac{n-p}{p(n-1)} < \beta < \frac{1}{q}\left(\frac{n}{n-1}\right).$$

If $p \geq n$ and $p \leq q < \infty$, then it suffices to choose any β, $0 < \beta < 1$ such that $\beta q < n/(n-1)$. In either case, β satisfies (8.31) and $\beta q < n/(n-1)$. Thus

$$\int_{T_{\tau,c,\rho}(\zeta)} V_1(z)^q \, d\lambda(z) \leq C \left[\int_{T_{\tau,c',\rho}(\zeta)} f(w)^p \, d\lambda(w)\right]^{q/p},$$

which by Lemma 8.17 with $\tau = n/\alpha$ goes to zero a.e. on S. Consequently, if $q \geq p$ and satisfies (8.30), then as above V_1 has T_τ-*limit* 0 in L^q a.e on S. If $q < p$, then by Hölder's inequality

$$\frac{1}{\nu(T_{\tau,c,\rho}(\zeta))} \int_{T_{\tau,c,\rho}(\zeta)} V_1(z)^q \, d\nu(z) \leq \left(\frac{1}{\nu(T_{\tau,c,\rho}(\zeta))} \int_{T_{\tau,c,\rho}(\zeta)} V_1(z)^p \, d\nu(z)\right)^{q/p}.$$

from which the result follows.   □

8.5. Related Results.

(1) Capacity Estimates of Exceptional Sets.

Our first remark involves capacity estimates of the exceptional sets of Theorem 8.16. As in [Co], if K is a compact subset of S, $0 < d \leq n$, the "non-isotropic" d-dimensional Hausdorff capacity of K is defined by

$$(8.32) \qquad H_d(K) = \inf \sum \delta_j^d,$$

where the infimum is over all covers $\{Q(\zeta_j, \delta_j)\}$ of K. If A is an arbitrary subset of S, then

$$(8.33) \qquad H_d(A) = \sup\{H_d(K) : K \text{ compact } \subset A\}.$$

Since $\sigma(Q(\zeta,\delta)) \approx \delta^n$, when $d = n$, H_n is equivalent to Lebesgue measure on S. When $n = 1$, the "non-isotropic" Hausdorff capacity corresponds to the usual Hausdorff capacity on the boundary. The main result of [Sto9] is the following:

Theorem 8.23. *Let f be a nonnegative measurable function on B satisfying (8.16) for some α, $0 < \alpha < n$, and some $p > n$. Then for each $\tau, 1 \leq \tau < n/\alpha$, there exists a set $E_\tau \subset S$ with $H_{\alpha\tau}(E_\tau) = 0$, such that G_f has T_τ-limit 0 at all $\zeta \in S \sim E_\tau$.*

The methods used to prove Theorem 8.23 also allowed us to prove the following generalization of a result due to G. T. Cargo [Car].

Theorem 8.24. *Let $\{a_k\}$ be a sequence in B satisfying*

$$(8.34) \qquad \sum_{1}^{\infty}(1 - |a_k|^2)^\alpha < \infty,$$

for some α, $0 < \alpha < n$, and let μ be the measure on B given by $\mu = \sum \delta_{a_k}$, where δ_{a_k} denotes the unit pointmass measure at a_k. Then for each τ, $1 \leq \tau < n/\alpha$, there exists a set $E_\tau \subset S$ with $H_{\alpha\tau}(E_\tau) = 0$, such that G_μ has T_τ-limit 0 at all $\zeta \in S \sim E_\tau$.

Remark. When $n = 1$, and μ is defined as above, then $G_\mu(z) = -\log|B(z)|$, where $B(z)$ is the Blaschke product with zeros at a_k. The result of Cargo [Car, Theorem 3] states that if $\{a_k\}$ is a Blaschke sequence satisfying (8.34) for some α, $0 < \alpha < 1$, then corresponding to each τ, $1 \leq \tau < 1/\alpha$ there is a set E_τ whose capacity of order $\alpha\tau$ is zero such that $B(z)$ has T_τ-limit of modulus one at each point of $S \sim E_\tau$. Since the result of Cargo was given in terms of the capacity of order $\beta\tau$ and not the Hausdorff capacity, Theorem 8.24 in the case $n = 1$ provides a slight improvement of that result.

(2) Weighted Limits of Potentials.

This area of investigation was motivated by an old result of M. Heins concerning weighted limits of $\log|B(z)|$, where $B(z)$ is a convergent Blaschke product in U. In [Hei] Heins proved that if B is a Blaschke product in U, then

$$(8.35) \qquad \liminf_{r \to 1}(1 - r)\log\frac{1}{|B(re^{i\theta})|} = 0$$

for all θ, $0 \leq \theta < 2\pi$. This result was extended to potentials on U as follows:

Theorem 8.25. *([Sto4]) If G_μ is the potential of a measure μ on U satisfying (8.2), then for all curves $\gamma : [0,1) \to U$ with $\lim_{t\to 1}\gamma(t) = 1$,*

$$(8.36) \qquad \liminf_{t \to 1}(1 - |\gamma(t)|)G_\mu(\gamma(t)e^{i\theta}) = 0,$$

for all θ, $0 \leq \theta < 2\pi$.

There are several extensions of Theorem 8.25 worth mentioning. The first involves the following question. Suppose E is a relatively closed subset of

B such that $t \in S$ is a limit point of E. What are necessarry and sufficient conditions on E such that

(8.37)
$$\liminf_{\substack{z \to t \\ z \in E}} (1 - |z|)^n G_\mu(z) = 0 ?$$

When $n = 1$, this question was answered by Luecking [Lu], and subsequently extended to the ball by Hahn and Singman [HSi] as follows: (8.37) holds for a set E if and only if the capacity of the sets

$$E \cap \{z \in B : |z - t| < \epsilon\}, \qquad \epsilon > 0$$

is bounded away from zero. Here, the capacity $c(K)$ of a compact set K can be defined as follows:

(8.38) $c(K) = \sup\{\mu(K) : \text{supp } \mu \subset K \text{ and } G_\mu \leq 1 \text{ on } K\}.$

For an arbitrary set A, the capacity of A is defined as the inner capacity, that is

$$c(A) = \sup\{c(K) : K \text{ is compact, } K \subset A\}.$$

In a different direction, one can show that (8.36) is equivalent to

(8.39)
$$\liminf_{r \to 1} (1 - r) M_\infty(G_\mu, r) = 0,$$

where

$$M_\infty(G_\mu, r) = \sup_{|z| = r} G_\mu(z).$$

This result was extended to euclidean potentials in the unit ball of $\mathbb{R}^n$ by S. Gardiner [Ga] and to invariant potentials on B by the author as follows:

Theorem 8.26. *([Sto5]) Let G_μ be the invariant potential of a measure μ on B satisfying (8.2).*

(a)   If $1 \leq p < (2n - 1)/2(n - 1)$, then
$$\lim_{r \to 1} (1 - r^2)^{n(1 - 1/p)} M_p(G_\mu, r) = 0.$$

(b)   If $n \geq 2$ and $(2n - 1)/2(n - 1) \leq p < (2n - 1)/(2n - 3)$, then
$$\liminf_{r \to 1} (1 - r^2)^{n(1 - 1/p)} M_p(G_\mu, r) = 0.$$

In the above, the p'th means $M_p(G_\mu, r)$ are as defined in Section 5.3. For related results involving absolutely continuous measures, the reader is referred to [Sto6].

9.
Gradient Estimates
and Riesz Potentials

In this chapter we present some recent results of K. T. Hahn, E. H. Yousffi, and the author [HSY] concerning invariant Riesz potentials and gradient estimates of invariant Green potentials in the unit ball B.

9.1. Gradient Estimates of Green Potentials.

The fundamental solution of the usual Laplacian in $\mathbb{R}^n$, $n > 2$, is given by

$$\Gamma(x, y) = c_n \, |x - y|^{2-n}$$

for an appropriate constant c_n. If one computes the gradient ∇ of Γ, one obtains

$$|\nabla_y \Gamma(x, y)| = (n - 2)c_n \, |x - y|^{1-n}.$$

The function $I_1(x) = |x|^{1-n}$ is called the Riesz kernel of order 1. More generally, the function $I_\alpha(x) = |x|^{\alpha - n}$, $0 < \alpha < n$, is called the Riesz kernel of order α. These functions have the property that

$$\mathcal{F}(I_\alpha)(y) = c_\alpha \, |y|^{-\alpha}$$

for an appropriate constant c_α, where $\mathcal{F}$ denotes the Fourier transform on $\mathbb{R}^n$. For $\alpha < n/2$, the Fourier transform is in the sense of distributions [Ste1, p. 117].

We now perform the same computation for the invariant Green's function G for the unit ball B.

Proposition 9.1. For fixed $w \in B$,

$$|\widetilde{\nabla}_z G(z, w)| = \frac{\sqrt{n+1}}{2n} \frac{(1 - |\varphi_z(w)|^2)^n}{|\varphi_z(w)|^{2n-1}}, \qquad z \neq w.$$

Proof. Since $G(z, w) = g(\varphi_w(z))$ and $\widetilde{\nabla}$ is $\mathcal{M}$-invariant,

$$|\widetilde{\nabla}_z G(z, w)| = |(\widetilde{\nabla} g)(\varphi_w(z))|.$$

Also, since g is $\mathcal{M}$-harmonic on $B \sim \{0\}$, by identity (3.13) and (3.9), for $z \neq 0$,

$$\begin{aligned}
|\widetilde{\nabla} g(z)|^2 &= \frac{1}{2} \widetilde{\Delta} g^2(z) = \frac{(1 - |z|^2)^2}{n+1} \, [g'(|z|)]^2 \\
&= \frac{n+1}{4n^2} \frac{(1 - |z|^2)^{2n}}{|z|^{4n-2}},
\end{aligned}$$

from which the result follows. $\square$

Definition. *The function*

$$(9.1) \qquad \mathcal{R}(z,w) = \frac{(1 - |\varphi_z(w)|^2)^n}{|\varphi_z(w)|^{2n-1}}, \qquad z \neq w$$

is called the **invariant Riesz kernel** *on B.*

If we set

$$h(z) = \frac{(1 - |z|^2)^n}{|z|^{2n-1}},$$

then $\mathcal{R}(z,w) = h(\varphi_z(w))$. By using the radial form of $\widetilde{\Delta}$, for $z \neq 0$,

$$\widetilde{\Delta}h(z) = \frac{(1 - |z|^2)^{n+1}}{(n+1)|z|^{2n+1}}[2n - 1 - |z|^2].$$

Thus $\widetilde{\Delta}h(z) > 0$ for all $z \in B \sim \{0\}$, and hence is $\mathcal{M}$-subharmonic on $B \sim \{0\}$. Therefore for fixed $w \in B$, $\mathcal{R}(z,w)$ is $\mathcal{M}$-subharmonic on $B \sim \{w\}$.

Definition. *For an appropriate measure μ, the function $\mathcal{R}\mu(z)$ defined by*

$$(9.2) \qquad \mathcal{R}\mu(z) = \int_B \mathcal{R}(z,w)\, d\mu(w)$$

is called the invariant **Riesz potential** *of the measure μ, provided of course that $\mathcal{R}\mu(z) < \infty$ almost everywhere on B. If μ is absolutely continuous with $du = f\, d\lambda$, then the Riesz potential of f, denoted by $\mathcal{R}f$, is defined by*

$$(9.3) \qquad \mathcal{R}f(z) = \int_B \mathcal{R}(z,w)f(w)\, d\lambda(w).$$

We will shortly prove that if μ is a regular measure satisfying

$$(9.4) \qquad \int_B (1 - |w|^2)^n\, d\mu(w) < \infty,$$

then $\mathcal{R}\mu(z) < \infty$ a.e. on B.

Note: Although we refer to $\mathcal{R}\mu$ as the Riesz potential of μ, this function will not be a potential in the sense of the definition given in Section 6.4. The function $\mathcal{R}\mu$ fails to be $\mathcal{M}$-superharmonic on B.

Definition. *If μ is a measure on B, the* **support** *of μ, denoted by $\operatorname{supp}\mu$ or S_μ is defined as the smallest relatively closed subset S_μ of B such that $\mu(B \sim S_\mu) = 0$ and $\mu(U \cap S_\mu) > 0$ for every open set U with $U \cap S_\mu \neq \phi$.*

Proposition 9.2. *Let μ be a nonnegative regular Borel measure on B satisfying (9.4). Then*

    (a) $\mathcal{R}\mu \in L^p_{loc}(B)$ *for all p, $0 < p < 2n/(2n-1)$.*

    (b) $\mathcal{R}\mu(z)$ *is a continuous $\mathcal{M}$-subharmonic function on $B \sim S_\mu$.*

Proof. Suppose μ satisfies (9.4). Define the functions U_1 and U_2 on B by

$$(9.5) \qquad U_1(z) = \int_{E(z)} \frac{1}{|\varphi_z(w)|^{2n-1}} \, d\mu(w),$$

$$(9.6) \qquad U_2(z) = \int_{B \sim E(z)} (1 - |\varphi_z(w)|^2)^n \, d\mu(w),$$

where as usual, $E(z) = \{w : |\varphi_z(w)| < \frac{1}{2}\}$. It is clear that there exist positive constants C_1 and C_2 such that

$$\mathcal{R}\mu(z) \le C_1 \, U_1(z) + C_2 \, U_2(z).$$

By identity (1.16),

$$U_2(z) \le \frac{2^n}{(1-|z|)^n} \int_B (1 - |w|^2) \, d\mu(w),$$

for all $z \in B$, and thus U_2 is bounded on any compact subset of B. Therefore $U_2 \in L^p_{loc}$ for all $p > 0$.

Consider the function $U_1(z)$. Let K be a compact subset of B. By compactness, there exists R, $0 < R < 1$, such that $E(z) \subset B_R$ for all $z \in K$. Thus by the continuous version of Minkowski's inequality, for $p \ge 1$,

$$\left[\int_K U_1^p(z) \, d\lambda(z) \right]^{1/p} \le \int_{B_R} \left[\int_K \chi_{E(w)}(z) |\varphi_z(w)|^{-p(2n-1)} \, d\lambda(z) \right]^{1/p} d\mu(w).$$

But

$$\int_K \chi_{E(w)}(z) |\varphi_z(w)|^{-p(2n-1)} \, d\lambda(z) \le \int_{E(w)} |\varphi_z(w)|^{-p(2n-1)} \, d\lambda(z)$$

$$= \int_{\frac{1}{2}B} |z|^{-p(2n-1)} \, d\lambda(z)$$

$$\le C \int_0^{\frac{1}{2}} r^{2n-1} r^{-p(2n-1)} \, dr < \infty$$

provided $p < 2n/(2n-1)$. Thus if $1 \le p < 2n/(2n-1)$,

$$\int_K U_1^p(z) \, d\lambda(z) \le C_p \, (\mu(B_R))^p,$$

where C_p is a constant depending only on p and n. For $0 < p < 1$, the result follows by Hölder's inequality. Thus $U_1 \in L^p_{loc}$ for all p, $0 < p < 2n/(2n-1)$, which proves (a).

Let S_μ denote the support of the measure μ, and let $a \in B \sim S_\mu$. Since $B \sim S_\mu$ is open, there exists $r_a > 0$ such that $\varphi_a(rB) \subset B \sim S_\mu$ for all r, $0 < r < r_a$. Thus since $z \to \mathcal{R}(z,w)$ is continuous and $\mathcal{M}$-subharmonic on $B \sim S_\mu$ for fixed $w \in S_\mu$, by Tonelli's theorem,

$$\int_S \mathcal{R}\mu(\varphi_a(rt))\, d\sigma(t) = \int_{S_\mu} \int_S \mathcal{R}(\varphi_a(rt), w)\, d\sigma(t)\, d\mu(w)$$

$$\geq \int_{S_\mu} \mathcal{R}(a, w)\, d\mu(w) = \mathcal{R}\mu(a)$$

for all r, $0 < r < r_a$. Thus $\mathcal{R}\mu$ is a continuous $\mathcal{M}$-subharmonic function on $B \sim S_\mu$. $\square$

We now state and prove the main result of this section:

Theorem 9.3. ([HSY])

(a) Let μ be a nonnegative regular Borel measure on B satisfying (9.4). Then $\widetilde{\nabla} G_\mu(z)$ exists a.e. on B, and

$$(9.7) \qquad |\widetilde{\nabla} G_\mu(z)| \leq \tfrac{\sqrt{n+1}}{2n} \mathcal{R}\mu(z) \qquad \text{a.e. on } B.$$

(b) If in addition μ is absolutely continuous with $d\mu = f\, d\lambda$, where $f \in L^q_{loc}$ for some $q > 2n$, then $\widetilde{\nabla} G_f(z)$ exists for all $z \in B$, and

$$(9.8) \qquad |\widetilde{\nabla} G_f(z)| \leq \tfrac{\sqrt{n+1}}{2n} \mathcal{R}f(z) \qquad \text{for all } z \in B.$$

Proof. Suppose μ is a regular Borel measure on B for which

$$(9.9) \qquad \frac{\partial G_\mu(z)}{\partial z_j} = \int_B \frac{\partial G(z,w)}{\partial z_j}\, d\mu(w)$$

holds either everywhere or a.e. on B. By definition of $\widetilde{\nabla}$ we have

$$|\widetilde{\nabla} G_\mu(z)|^2 = \langle \widetilde{\nabla} G_\mu(z), \widetilde{\nabla} G_\mu(z) \rangle = 4 \sum_{i,j} b^{ij}(z) \overline{\frac{\partial G_\mu(z)}{\partial z_i}} \frac{\partial G_\mu(z)}{\partial z_j}.$$

Thus by (9.9) and linearity

$$|\widetilde{\nabla} G_\mu(z)|^2 = \int_B \int_B \langle \widetilde{\nabla}_z G(z,w), \widetilde{\nabla}_z G(z,\zeta) \rangle\, d\mu(w)\, d\mu(\zeta).$$

Since $\langle\,,\,\rangle$ defines an inner product which satisfies $|\langle\tilde\nabla f,\tilde\nabla g\rangle| \le |\tilde\nabla f|\,|\tilde\nabla g|$, we have

$$|\tilde\nabla G_\mu(z)|^2 \le \int_B \int_B |\tilde\nabla_z G(z,w)||\tilde\nabla_z G(z,\zeta)|\,d\mu(w)\,d\mu(\zeta)$$
$$= \left[\int_B |\tilde\nabla_z G(z,w)|\,d\mu(w)\right]^2,$$

which when combined with Proposition 9.1 proves (9.7) or (9.8). Thus to complete the proof we need to show identity (9.9) holds either everywhere or almost everywhere whenever μ satisfies the hypothesis.

We first consider $\dfrac{\partial G(z,w)}{\partial z_j}$. By (3.16), for real valued u,

(9.10)
$$|\tilde\nabla u|^2 = \frac{4}{n+1}(1-|z|^2)[|\partial u|^2 - |\langle z,\partial u\rangle|^2]$$
$$\ge \frac{4}{n+1}(1-|z|^2)^2\,|\partial u|^2.$$

Therefore by Proposition 9.1 and the above,

$$\left|\frac{\partial G(z,w)}{\partial z_j}\right| \le |\partial_z G(z,w)| \le \frac{\sqrt{n+1}}{2(1-|z|^2)}|\tilde\nabla_z G(z,w)| = \frac{(n+1)}{4n}\frac{\mathcal{R}(z,w)}{(1-|z|^2)}.$$

By the $\mathcal{M}$-invariance of λ and integration in polar coordinates,

$$\int_{E(z)} \mathcal{R}^p(z,w)\,d\lambda(w) = \int_{|\zeta|<\frac12} \frac{(1-|\zeta|^2)^{pn}}{|\zeta|^{p(2n-1)}}\,d\lambda(\zeta)$$
$$\le C \int_0^{\frac12} r^{2n-p(2n-1)-1}\,dr < \infty,$$

provided $p < 2n/(2n-1)$. Since $\mathcal{R}$ is a radial function of $\varphi_z(w)$, the above result is still valid when the integration is performed with respect to z. Thus $\dfrac{\partial G(z,w)}{\partial z_j}$ is locally p-integrable in either variable for all $p < 2n/(2n-1)$.

Let U be a relatively compact open subset of B. Choose a relatively compact open subset V of B such that $\overline{U} \subset V$. Set $\mu_1 = \mu_{|V}$ and $\mu_2 = \mu - \mu_1$. Since $\mu = \mu_1 + \mu_2$,

$$G_\mu(z) = G_{\mu_1}(z) + G_{\mu_2}(z).$$

Consider $G_{\mu_2}(z)$ for $z \in U$. We first note that by the definition of $\mathcal{R}(z,w)$ there exists a constant C (depending on V) such that

$$\mathcal{R}(z,w) \le C\,(1-|w|^2)^n$$

for all $z \in U$, $w \in B \sim V$. Thus for all $z \in U$

$$\int_{B \sim V} \left| \frac{\partial G(z,w)}{\partial z_j} \right| d\mu(w) \leq C \int_{B \sim V} (1 - |w|^2)^n \, d\mu(w),$$

which is finite by hypothesis. Thus since $z \to G(z,w)$ is C^∞ on U for all $w \in B \sim V$, we obtain

$$\frac{\partial G_{\mu_2}(z)}{\partial z_j} = \int_{B \sim V} \frac{\partial G(z,w)}{\partial z_j} \, d\mu(w) \qquad \text{for all} \quad z \in U.$$

We now consider G_{μ_1}. For $j = 1, ..., n$, set

$$F_j(z) = \int_V \frac{\partial G(z,w)}{\partial z_j} \, d\mu(w).$$

Since $\dfrac{\partial G(z,w)}{\partial z_j}$ is locally integrable as a function of z, by Fubini's theorem

$$\int_K |F_j(z)| \, d\lambda(z) < \infty$$

for every compact subset K of U. Thus $|F_j(z)| < \infty$ for a.e. $z \in U$. Furthermore, for all $z \in U$, the function $w \to F_j(z + w)$ is locally integrable in a neighborhood of $0 \in \mathbb{C}^n$. Therefore, if we let e_j denote the unit vector with a 1 in the j'th coordinate, by Fubini's theorem, the function $\zeta \to F_j(z + \zeta e_j)$ is locally integrable near $0 \in \mathbb{C}$ for a.e. $z \in U$.

Suppose $d\mu = f \, d\lambda$ where $f \in L^q_{loc}(B)$ for some $q > 2n$. Let $p = q/(q-1)$ denote the conjugate exponent of q. Then $p < 2n/(2n-1)$. Thus by Hölder's inequality,

$$|F_j(z)| \leq \left[\int_V \left| \frac{\partial G(z,w)}{\partial z_j} \right|^p d\lambda(w) \right]^{1/p} \left[\int_V f^q(w) \, d\lambda(w) \right]^{1/q} < \infty.$$

Hence in this case, $F_j(z)$ is finite for all $z \in U$. Furthermore, since $\dfrac{\partial G(z,w)}{\partial z_j}$ is locally p-integrable and continuous for $w \neq z$, it is easy to show that $F_j(z)$ is continuous on U.

Finally, for $z \in U$ and $\zeta \in \mathbb{C}$ sufficiently close to 0, consider $\gamma(t) = z + t \zeta e_j$. Since

$$\frac{d}{dt} G(\gamma(t), w) = 2 \text{Re} \left[\zeta \frac{\partial G(\gamma(t), w)}{\partial z_j} \right],$$

we have

$$G_{\mu_1}(z + \zeta e_j) - G_{\mu_1}(z) = \int_V \left(\int_0^1 \frac{d}{dt} G(\gamma(t), w) dt \right) d\mu(w)$$

$$= 2\text{Re} \left[\zeta \int_0^1 F_j(z + t\zeta e_j) dt \right].$$

This identity holds a.e. on U, and everywhere in the case where $f \in L^q_{loc}$, $q > 2n$. Also,

$$\lim_{t \to 0} \int_0^1 F_j(z + t\zeta e_j) dt = F_j(z)$$

a.e. in the case where F_j is integrable, and everywhere when F_j is continuous. Finally, since $\frac{\partial}{\partial z_j} = \frac{1}{2}(\frac{\partial}{\partial x_j} - i \frac{\partial}{\partial y_j})$, by taking ζ to be first real and then purely imaginary, we obtain

$$\frac{\partial G_{\mu_1}(z)}{\partial z_j} = F_j(z)$$

a.e. on U, and everywhere if $f \in L^q_{loc}$ for some $q > 2n$. Combining this with the result for G_{μ_2} shows that (9.7) holds a.e. on B if μ satisfies (9.4), and everywhere on B if μ satisfies the additional hypothesis $d\mu = f \, d\lambda$ with $f \in L^q_{loc}(B)$ for some $q > 2n$. $\square$

9.2. L^p Inequalities for the Riesz Operator.

It is well known (see [Ste1, p.119]) that for $p > 1$, the Riesz operator I_α satisfies

$$\|I_\alpha(f)\|_q \leq C \|f\|_p,$$

where $\frac{1}{q} = \frac{1}{p} - \frac{\alpha}{n}$, and

$$I_\alpha(f) = \int_{\mathbb{P}^n} \frac{f(y)}{|x - y|^{n-\alpha}} \, dy.$$

We now prove an analogous result for the invariant Riesz operator. Rather than proving the result for just the operator $\mathcal{R}$, it will prove useful to consider the more general operator $\mathcal{R}_{\alpha,\beta}$ which is defined as follows:

Definition. *Let f be a measurable function on B. For $\alpha, \beta > 0$, the invariant Riesz operator $\mathcal{R}_{\alpha,\beta}$ of order α, β is defined by*

$$(9.11) \qquad \mathcal{R}_{\alpha,\beta} f(z) = \int_B \frac{(1 - |\varphi_z(w)|^2)^\beta}{|\varphi_z(w)|^{2n-\alpha}} f(w) \, d\lambda(w),$$

whenever the integral exists. The function

$$\mathcal{R}_{\alpha,\beta}(z, w) = \frac{(1 - |\varphi_z(w)|^2)^\beta}{|\varphi_z(w)|^{2n-\alpha}}$$

*is called the **invariant Riesz kernel** or order α, β.*

The case $\alpha = 1$ and $\beta = n$ gives the Riesz kernel as introduced in the previous section. Also, as in the previous section, one can show that

$$\mathcal{R}_{\alpha,\beta}(z,w) = \sqrt{n+1}|\tilde{\nabla}_z G_{\alpha,\beta}(z,w)|,$$

where $G_{\alpha,\beta}(z,w)$ is defined by

$$G_{\alpha,\beta}(z,w) = \int_{|\varphi_z(w)|}^1 \frac{(1-t^2)^{\beta-1}}{t^{2n-\alpha}}\, dt.$$

Furthermore, by computation, one can also show that for $0 < \beta \le n$ and $2 \le \alpha \le 2n$, the function $z \to \mathcal{R}_{\alpha,\beta}(z,w)$ is $\mathcal{M}$-superharmonic on B for fixed $w \in B$.

Theorem 9.4. *([HSY]) Let $\alpha > 0, \beta > 0$ and $1 \le p < q < +\infty$ be such that*

(9.12a)
$$\frac{1}{q} + \frac{\alpha}{2n} = \frac{1}{p},$$

(9.12b)
$$\beta + \frac{\alpha}{2} \ge n.$$

(1) *If $p > 1$, then the Riesz operator $\mathcal{R}_{\alpha,\beta}$ is bounded from $L^p(B,\lambda)$ into $L^q(B,\lambda)$.*
(2) *If $p = 1$, then the Riesz operator $\mathcal{R}_{\alpha,\beta}$ is of weak-type $(1,q)$ from $L^q(B,\lambda)$ into $L^1(B,\lambda)$, that is, there exists a positive constant C such that*

$$\lambda(\{z : |(\mathcal{R}_{\alpha,\beta}f)(z)| > t\}) \le \left(\frac{C\|f\|_1}{t}\right)^q,$$

for all $t > 0$ and $f \in L^1(B,\lambda)$.

Remark: It is easily shown that inequality (9.12b) is equivalent to

$$\frac{\beta}{2n} + \frac{1}{2p} + \frac{(\beta-n)}{\alpha q} - \frac{1}{2} \ge 0.$$

Furthermore, since $\frac{1}{2}\alpha = n\left(\frac{1}{p} - \frac{1}{q}\right)$, we also have

$$\beta \ge n\left(1 - \frac{1}{p}\right) + \frac{n}{q}.$$

When $\beta = n$, (9.12b) is satisfied for all $\alpha > 0$.

Proof. We first show that $\mathcal{R}_{\alpha,\beta}$ is of weak type (p,q) for all p, q, α, β satisfying the hypothesis. Note that $\mathcal{R}_{\alpha,\beta}f = f * h$, where

$$h(z) = \frac{(1 - |z|^2)^\beta}{|z|^{2n-\alpha}}, \quad z \in B.$$

For $0 < r < 1$, write h in the form $h(z) = h_0 + h_1$, where $h_0 = \chi_{rB}h$. Here χ_E denotes the characteristic function of the set E. Without loss of generality assume that f is nonnegative. Since

$$\mathcal{R}_{\alpha,\beta}f = f * h_0 + f * h_1,$$

we obtain that for for $t > 0$,

(9.13)
$$\lambda(\{z : (\mathcal{R}_{\alpha,\beta}f)(z) > t\}) \le \lambda(\{z : (f * h_0)(z) > \tfrac{1}{2}t\})$$
$$+ \lambda(\{z : (f * h_1)(z) > \tfrac{1}{2}t\}).$$

If $\|f\|_p = 1$, then Lemma 4.4 implies that

(9.14)
$$\lambda(\{z : (f * h_0)(z) > \tfrac{1}{2}t\}) \le \left(\frac{2\|f * h_0\|_p}{t}\right)^p \le C\left(\frac{\|h_0\|_1}{t}\right)^p.$$

Since

$$\|h_0\|_1 = \int_{rB} \frac{(1 - |w|^2)^\beta}{|w|^{2n-\alpha}}\, d\lambda(w) = 2n \int_0^r x^{\alpha-1}(1 - x^2)^{\beta-n-1}\, dx,$$

l'Hospital's rule implies that there exists a positive constant C_1 such that

(9.15)
$$\|h_0\|_1 \le C_1 \begin{cases} r^\alpha, & \beta > n, \\ r^\alpha \log \dfrac{e}{(1 - r^2)}, & \beta = n, \\ r^\alpha(1 - r^2)^{\beta-n}, & \beta < n. \end{cases}$$

Consider $f * h_1$. Assume $p > 1$ and let $p' = p/(p - 1)$. Then by Lemma 4.4,

$$\|f * h_1\|_\infty^{p'} \le \|f\|_p^{p'}\|h_1\|_{p'}^{p'} = \|h_1\|_{p'}^{p'}$$
$$= \int_{B \setminus rB} \frac{(1 - |w|^2)^{\beta p'}}{|w|^{p'(2n-\alpha)}}\, d\lambda(w)$$
$$= 2n \int_r^1 x^{2n-1-(2n-\alpha)p'}(1 - x^2)^{\beta p'-n-1}\, dx$$
$$= 2n \int_r^1 x^{\frac{\alpha p - 2n}{p-1} - 1}(1 - x^2)^{\frac{(\beta - n)p + n}{p-1} - 1}\, dx.$$

Since α and β satisfy (9.12a) and (9.12b) respectively, we have $(\alpha p - 2n)/(p-1) < 0$ and $[(\beta - n)p + n]/(p-1) > 0$. Hence by l'Hospital's rule, there exists $C_2 > 0$ such that for all r, $0 < r < 1$,

$$\int_r^1 x^{\frac{\alpha p - 2n}{p-1}-1}(1-x^2)^{\frac{(\beta-n)p+n}{p-1}-1} \, dx \le C_2 \, r^{\frac{\alpha p - 2n}{p-1}}(1-r^2)^{\frac{(\beta-n)p+n}{p-1}}.$$

Therefore,

$$(9.16) \quad \|f * h_1\|_\infty \le C_2 \, r^{\frac{\alpha p - 2n}{p}}(1-r^2)^{\frac{(\beta-n)p+n}{p}} = C_2 \left(r(1-r^2)^{-\gamma q}\right)^{-\frac{2n}{q}},$$

where

$$\gamma = \frac{\beta}{2n} + \frac{1}{2p} - \frac{1}{2},$$

which is easily seen to be positive. If $p = 1$, then by (9.12a) $\alpha/2n < 1$. Thus

$$\|f * h_1\|_\infty \le \|f\|_1 \|h_1\|_\infty \le r^{\alpha - 2n}(1-r^2)^\beta.$$

Therefore (9.16) holds for all $p \ge 1$.

Let $t > 0$ be given. Choose r, $0 < r < 1$, so that

$$(9.17) \qquad t = 2C_2 \left(r(1-r^2)^{-\gamma q}\right)^{-\frac{2n}{q}}.$$

As a consequence of (9.16), we see that

$$(9.18) \qquad \lambda(\{z : (f * h_1)(z) > \tfrac{1}{2}t\}) = 0.$$

Suppose $\beta < n$. Since $\gamma + \frac{\beta-n}{\alpha q} \ge 0$ by (9.12b), it follows from (9.17) that

$$r^\alpha(1-r^2)^{\beta-n} = C \, t^{-\frac{\alpha q}{2n}}(1-r^2)^{\alpha q(\gamma + \frac{\beta-n}{\alpha q})} \le C \, t^{-\frac{\alpha q}{2n}}.$$

Since $\gamma > 0$, if $\beta > n$, then

$$r^\alpha = C \, t^{-\frac{\alpha q}{2n}}(1-r^2)^{\gamma q} \le C \, t^{-\frac{\alpha q}{2n}},$$

and for $\beta = n$,

$$r^\alpha \log \frac{e}{(1-r^2)} = C \, t^{-\frac{\alpha q}{2n}}(1-r^2)^{\gamma q} \log \frac{e}{(1-r^2)} \le C \, t^{-\frac{\alpha q}{2n}}.$$

Thus for all β, by (9.15) and the above,

$$\|h_0\|_1 \le C \, t^{-\frac{\alpha q}{2n}}.$$

Finally, since $-\frac{\alpha q p}{2n} - p = -q$, combining (9.13), (9.14), (9.18), and the above, we obtain

$$\lambda(\{z : |(\mathcal{R}_{\alpha,\beta} f)(z)| > t\}) \leq \left(\frac{C\|f\|_p}{t}\right)^q,$$

for all $t > 0$ and $f \in L^p(B, \lambda)$, where C is a positive constant. Thus $\mathcal{R}_{\alpha,\beta}$ is of weak type (p, q) for all $1 \leq p < q$ satisfying (9.12a) and (9.12b). In particular, by (9.12a) $\mathcal{R}_{\alpha,\beta}$ is of weak type $(1, \frac{2n}{2n-\alpha})$.

Now suppose that $1 < p < q$, and α, β satisfying (9.12a) and (9.12b) are fixed. Choose $p_1 > p$ and $q_1 > q$ satisfying

$$\frac{1}{q_1} + \frac{\alpha}{2n} = \frac{1}{p_1}$$

for the given α. Applying the above to (p_1, q_1) in place of (p, q), we see that $\mathcal{R}_{\alpha,\beta}$ is a sublinear transformation on $L^1(B, \lambda) + L^{p_1}(B, \lambda)$ which is of weak types (p_1, q_1) and $(1, \frac{2n}{2n-\alpha}) = (p_0, q_0)$. The Marcinkiewicz interpolation theorem ([ZY]) implies that $\mathcal{R}_{\alpha,\beta}$ is bounded from $L^{p_\theta}(B, \lambda)$ into $L^{q_\theta}(B, \lambda)$, where for $0 < \theta < 1$,

$$\frac{1}{p_\theta} = \frac{1-\theta}{p_0} + \frac{\theta}{p_1}, \quad \text{and} \quad \frac{1}{q_\theta} = \frac{1-\theta}{q_0} + \frac{\theta}{q_1}.$$

Take $\theta = \frac{p_1(p-1)}{p(p_1-1)}$. Since $p_1 > p$, we have $0 < \theta < 1$, and $p_\theta = p$. Using (9.12a), it is also easy to check $q_\theta = q$. Thus $\mathcal{R}_{\alpha,\beta}$ is bounded from $L^p(B, \lambda)$ into $L^q(B, \lambda)$. □

The following theorem extends Theorem 9.4 to include the case of weighted L^p norms.

Theorem 9.5. *Suppose that $\alpha > 0$ and $1 < p < q < +\infty$ satisfy*

(9.19)
$$\frac{1}{p} < \frac{1}{q} + \frac{\alpha}{2n}.$$

If $s \in \mathbb{R}$ satisfies

(9.20)
$$\frac{n}{q} - n < s < n\left(\frac{1}{p} - \frac{1}{q}\right),$$

then there is a positive constant $C = C(\alpha, p, q, s)$ such that

$$\|(1 - |z|^2)^s (\mathcal{R}_{\alpha,n} f)\|_q \leq C \|(1 - |z|^2)^s f\|_p,$$

for all measurable functions f such that $(1 - |z|^2)^s f(z) \in L^p(B, \lambda)$.

Remark: Although we stated the above result only for the case $\beta = n$, the analogous result is still valid for the operator $\mathcal{R}_{\alpha,\beta}$ whenever s and β satisfy

$$\frac{n}{q} - \beta < s < \beta + n\left(\frac{1}{p} - \frac{1}{q} - 1\right),$$

Proof. By virtue of (9.19) there exits $0 < \delta < \alpha$ satisfying

(9.21)
$$\frac{1}{q} + \frac{\delta}{2n} = \frac{1}{p}.$$

It follows from (9.20) that there exists $r \in (1, p)$ such that

$$n\left(\frac{1}{r} - \frac{1}{p}\right) + n\left(\frac{1}{q} - 1\right) = \frac{n}{r} - n - n\left(\frac{1}{p} - \frac{1}{q}\right) < s < n\left(\frac{1}{p} - \frac{1}{q}\right).$$

By equality (9.21) we have $\frac{1}{2}\delta = n\left(\frac{1}{p} - \frac{1}{q}\right)$. Thus by the above,

$$\frac{n}{r} - n - \frac{\delta}{2} < s < \frac{\delta}{2}.$$

As a consequence,

$$\left(s + n\left(1 - \frac{1}{r}\right) + \frac{\delta}{2}\right) - 2s > n - \frac{n}{r}, \qquad \text{and}$$

$$s + n\left(1 - \frac{1}{r}\right) + \frac{\delta}{2} > 0.$$

Thus we can choose a real number γ satisfying

(9.22a)
$$0 < \gamma < s + n\left(1 - \frac{1}{r}\right) + \frac{\delta}{2}$$

(9.22b)
$$\gamma - 2s > n\left(1 - \frac{1}{r}\right)$$

Let $\alpha' = r\delta$ and $\beta = r(n + s - \gamma)$. Then by (9.21) and (9.22a),

$$\frac{r}{q} + \frac{\alpha'}{2n} = \frac{r}{p} \qquad \text{and} \qquad \beta + \frac{\alpha'}{2} > n.$$

Thus by Theorem 9.4 we see that there is a constant $C = C(p, q, s, \alpha, \beta) > 0$ such that

(9.23)
$$\|\mathcal{R}_{\alpha', \beta} h\|_{q/r} \leq C \|h\|_{p/r},$$

for all $h \in L^{p/r}(B, \lambda)$.

Now let $f : B \to [0, \infty)$ be a measurable function. By the definition of $\mathcal{R}_{\alpha, n}$ and the $\mathcal{M}$-invariance of λ, we have

$$\mathcal{R}_{\alpha, n} f(z) = \int_B \frac{(1 - |w|^2)^n}{|w|^{2n - \alpha}} (1 - |\varphi_z(w)|^2)^{-s} (h \circ \varphi_z)(w) \, d\lambda(w),$$

where

$$h(z) = (1 - |z|^2)^s f(z), \quad z \in B.$$

By (1.16)

$$1 - |z|^2 = \frac{(1 - |\varphi_z(w)|^2)|1 - \langle z, w \rangle|^2}{1 - |w|^2}.$$

Therefore

$$(1 - |z|^2)^s \, \mathcal{R}_{\alpha,n} f(z) = \int_B \frac{(1 - |w|^2)^{n-s}}{|w|^{2n-\alpha}} |1 - \langle z, w \rangle|^{2s} (h \circ \varphi_z)(w) \, d\lambda(w).$$

Let $r' = r/(r-1)$ denote the conjugate exponent of r. Then by Hölder's inequality,

$$(1 - |z|^2)^s \, (\mathcal{R}_{\alpha,n} f)(z)$$

$$\leq \left(\int_B \frac{(1 - |w|^2)^{r(n+s-\gamma)}}{|w|^{2n-r\delta}} (h \circ \varphi_z)^r(w) \, d\lambda(w) \right)^{1/r}$$

$$\times \left(\int_B \frac{(1 - |w|^2)^{(\gamma-2s)r'}}{|w|^{2n-(\alpha-\delta)r'}} |1 - \langle z, w \rangle|^{2sr'} \, d\lambda(w) \right)^{1/r'}.$$

By (9.22a) and (9.22b), $\gamma > 0$ and $(\gamma - 2s)r' > n$. Thus by Proposition 8.18,

$$M = \sup_{z \in B} \left(\int_B \frac{(1 - |w|^2)^{(\gamma-2s)r'}}{|w|^{2n-(\alpha-\delta)r'}} |1 - \langle z, w \rangle|^{2sr'} \, d\lambda(w) \right)^{1/r'} < \infty.$$

As a consequence,

$$(1 - |z|^2)^s \, (\mathcal{R}_{\alpha,n} f)(z)$$

$$\leq M \left(\int_B \frac{(1 - |w|^2)^\beta}{|w|^{2n-r\delta}} (h \circ \varphi_z)^r(w) \, d\lambda(w) \right)^{1/r}$$

$$= M \left((\mathcal{R}_{\alpha',\beta} h^r)(z) \right)^{1/r}.$$

Thus by inequality (9.23) and the above,

$$\|(1 - |z|^2)^s \mathcal{R}_{\alpha,n} f\|_q \leq M \left(\|\mathcal{R}_{\alpha',\beta} h^r\|_{q/r} \right)^{1/r}$$

$$\leq C M \left(\|h^r\|_{p/r} \right)^{1/r} = C M \|(1 - |z|^2)^s f\|_p,$$

which proves the result. $\square$

As a corollary to the previous theorem we obtain the following:

Corollary 9.6. *Suppose $1 < p < \infty$.*

(a) If $\alpha > 0$, then for all $q > p$ satisfying (9.19), there exists a positive constant C such that

$$\|\mathcal{R}_{\alpha,n}f\|_q \le C \|f\|_p \qquad \text{for all} \quad f \in L^p(B,\lambda).$$

(b) For all q satisfying

$$(9.24) \qquad p < q < \begin{cases} \dfrac{2np}{2n-p}, & p < 2n, \\[2mm] \infty, & p \ge 2n, \end{cases}$$

there exists a positive constant C such that

$$\|\tilde{\nabla}G_f\|_q \le C \|f\|_p \qquad \text{for all} \quad f \in L^p(B,\lambda).$$

Proof. Part (a) follows from the previous theorem by taking $s = 0$. If q satisfies (9.24), then

$$\frac{1}{p} < \frac{1}{q} + \frac{1}{2n}.$$

Part (b) now follows immediately by (a) and Theorem 9.3. If $f \in L^p(B,\lambda)$ for some $p > 1$, then by Hölder's inequality f satisfies the hypothesis of Theorem 9.3, namely that $\int_B (1 - |w|^2)^n f(w) d\lambda(w) < \infty$.

We now show that for suitable β, $\mathcal{R}_{\alpha,\beta}$ ($\alpha > 0$) is also a bounded mapping of $L^p(B,\lambda)$ into $L^p(B,\lambda)$. For the proof of this result, we need the following well know result usually referred to as Shur's theorem.

Lemma 9.7. *Let (X,μ) be a measure space, $1 < p < \infty$, and K a nonnegative measurable functions on $X \times X$. If there exists a positive measurable function h on X and positive constants A and B such that*

$$\int_X K(x,y)h(x)^p \, d\mu(x) \le A\, h(y)^p$$

for μ - a.e. $y \in X$, and

$$\int_X K(x,y)h(y)^q \, d\mu(y) \le B h(x)^q$$

for μ - a.e. $x \in X$, where $\frac{1}{q} + \frac{1}{p} = 1$, then T defined by

$$Tf(x) = \int_X K(x,y)f(y) \, d\mu(y)$$

is a bounded operator on $L^p(X,\mu)$.

The proof of the lemma follows directly from Hölder's inequality and Fubini's theorem, and thus is left as an exercise for the interested reader. If the function K is symmetric, i.e., $K(x,y) = K(y,x)$, then by symmetry, T is also bounded on $L^q(X,\mu)$, and hence on $L^r(X,\mu)$ for all r between p and q.

Theorem 9.8. *Let $1 < p < \infty$, and let $\alpha > 0$.*

(1) *Then for all $\beta > n \max\{\frac{1}{p}, 1 - \frac{1}{p}\}$, $\mathcal{R}_{\alpha,\beta}$ is a bounded operator on $L^p(B, \lambda)$.*

(2) *Furthermore, for all s satisfying*

$$(9.25) \qquad 0 < s < n \min\left\{\frac{1}{p}, 1 - \frac{1}{p}\right\},$$

$$\|(1 - |z|^2)^s \mathcal{R}_{\alpha,n} f\|_p \leq C \, \|(1 - |z|^2)^s f\|_p,$$

for all measurable functions f such that $(1 - |z|^2)^s f(z) \in L^p(B, \lambda)$.

Proof. Let $q = p/(p-1)$ be the conjugate exponent of p. To prove the result we show the existence of a function $g : B \to (0, \infty)$ satisfying

$$(9.26a) \qquad \int_B R_{\alpha,\beta}(z, w)\, g(w)^q \, d\lambda(w) \leq A\, g(z)^q,$$

$$(9.26b) \qquad \int_B R_{\alpha,\beta}(z, w)\, g(z)^p \, d\lambda(z) \leq B\, g(w)^p,$$

for positive constants A and B. By the previous lemma, the existence of such a g implies that $\mathcal{R}_{\alpha,\beta}$ is a bounded operator on $L^p(B, \lambda)$ (and also on $L^q(B, \lambda)$).

Without loss of generality we assume $\max\{\frac{1}{p}, \frac{1}{q}\} = \frac{1}{q}$. Choose γ such that

$$\frac{n}{pq} < \gamma < \frac{\beta}{p}.$$

By the hypothesis, such a choice is possible. Let $g(z) = (1 - |z|^2)^\gamma$, and consider

$$\int_B R_{\alpha,\beta}(z, w) g(w)^q \, d\lambda(w) = \int_{E(z)} R_{\alpha,\beta}(z, w) g(w)^q \, d\lambda(w)$$

$$+ \int_{B \sim E(z)} R_{\alpha,\beta}(z, w) g(w)^q \, d\lambda(w).$$

Since $(1 - |w|^2) \approx (1 - |z|^2)$ for all $w \in E(z)$,

$$\int_{E(z)} R_{\alpha,\beta}(z, w)\, g(w)^q \, d\lambda(w) \leq C \, (1 - |z|^2)^{q\gamma} \int_{E(z)} |\varphi_z(w)|^{\alpha - 2n} \, d\lambda(w)$$

$$\leq C \, (1 - |z|^2)^{q\gamma}.$$

Since $|\varphi_z(w)| \geq \frac{1}{2}$ for all $w \in B \sim E(z)$, by (1.16)

$$\int_{B \sim E(z)} R_{\alpha,\beta}(z,w) \, g(w)^q \, d\lambda(w) \leq C \, (1 - |z|^2)^\beta \int_B \frac{(1 - |w|^2)^{\beta + q\gamma}}{|1 - \langle z, w \rangle|^{2\beta}} \, d\lambda(w).$$

Since $\beta - q\gamma > 0$ and $\beta + q\gamma > n$ (which is easily seen to be the case for the above choice of γ), by Proposition 8.18,

$$\int_B \frac{(1 - |w|^2)^{\beta + q\gamma}}{|1 - \langle z, w \rangle|^{2\beta}} \, d\lambda(w) \leq C \, (1 - |z|^2)^{q\gamma - \beta},$$

which proves (9.26a). Since $\beta - p\gamma > 0$ and $\beta + p\gamma > n$ also hold for the above choice of γ, reversing the roles of z and w gives (9.26b).

For the proof of (2), choose $\epsilon > 0$ such that

$$n - s > \epsilon + \frac{n}{q}.$$

This is possible since by hypothesis $n - s > n \max\{\frac{1}{p}, \frac{1}{q}\} = \frac{n}{q}$. Set $\beta = n/q + \epsilon$. Then $\beta > n \max\{\frac{1}{p}, \frac{1}{q}\}$. As in Theorem 9.5,

$$(1 - |z|^2)^s \mathcal{R}_{\alpha,n} f(z) = \int_B \frac{(1 - |w|^2)^{n-s}}{|w|^{2n-\alpha}} |1 - \langle z, w \rangle|^{2s} h(\varphi_z(w)) \, d\lambda(w)$$

$$\leq C \int_B (1 - |w|^2)^{(n-s-\beta)} \frac{(1 - |w|^2)^\beta}{|w|^{2n-\alpha}} h(\varphi_z(w)) \, d\lambda(w)$$

$$\leq C \, (\mathcal{R}_{\alpha,\beta} h)(z),$$

where $h(w) = (1 - |w|^2)^s f(w)$. The above inequality holds provided $n - s - \beta \geq 0$, which is the case for the above choice of β. Thus by part (1),

$$\|(1 - |z|^2)^s \mathcal{R}_{\alpha,n} f\|_p \leq C \, \|\mathcal{R}_{\alpha,\beta} h\|_p \leq C \, \|h\|_p = C \, \|(1 - |z|^2)^s f\|_p. \quad \square$$

Note: As a consequence of Theorem 9.8, $\mathcal{R}_{\alpha,n}$ ($\alpha > 0$) is also a bounded operator on $L^p(B, \lambda)$ for all $p > 1$. Therefore, Corollary 9.6 is also valid for the case $q = p$.

10.
Spaces of Invariant
Harmonic Functions

In this chapter we take a brief look at weighted Bergman and Dirichlet type spaces of $\mathcal{M}$-harmonic functions on B. These are motivated by the classical Bergman and Dirichlet spaces of holomorphic functions on the unit disc U.

In Section 10.1 we derive several useful invariant mean value inequalities for $|h|^p$ and $|\tilde{\nabla}h|^p$, $0 < p < \infty$, for $\mathcal{M}$-harmonic functions h on B. Using these results, we prove a generalization of a theorem of Hardy and Littlewood on the comparison of the rate of growth of the means $M_p(h, r)$ and $M_p(\tilde{\nabla}h, r)$ as $r \to 1$. Finally, in section 10.3 we introduce the weighted Bergman and Dirichlet spaces of $\mathcal{M}$-harmonic functions on B, and prove some basic results about these spaces.

10.1. Mean Value Inequalities for $|h|^p$ and $|\tilde{\nabla}h|^p$, $0 < p < \infty$.

By inequality (4.3), if f is $\mathcal{M}$-subharmonic on B, then

$$f(a) \le \frac{1}{\lambda(E(a,r))} \int_{E(a,r)} f(w)\, d\lambda(w),$$

for all $a \in B$, $0 < r < 1$, with equality if f is $\mathcal{M}$-harmonic. The following proposition extends this inequality to f^p, $(f \ge 0)$ for all values of p, $0 < p < \infty$.

Proposition 10.1. *Let f be a nonnegative $\mathcal{M}$-subharmonic function on B. Then for all p, $0 < p < \infty$, $a \in B$, and $0 < r < 1$,*

$$f^p(a) \le \frac{C(n,p,r)}{r^{2n}} \int_{E(a,r)} f^p(w)\, d\lambda(w)$$

where

(10.1)
$$C(n,p,r) = \begin{cases} (1-r^2)^n, & 1 \le p < \infty, \\ 2^{2n/p}, & 0 < p < 1. \end{cases}$$

Proof. For $p \ge 1$, the result is an immediate consequence of Jensen's inequality. For the case $0 < p < 1$, we use a variation of the argument given in [Pa, Lemma 2.4]. By (4.4), $\lambda(E(a,r)) = r^{2n}/(1-r^2)^n$. Therefore,

$$f(a) \le \frac{1}{r^{2n}} \int_{E(a,r)} f(w)\, d\lambda(w).$$

for all $a \in B$, $0 < \epsilon < 1$. Fix r, $0 < r < 1$, and let

$$C_p^p = \int_{E(a,r)} f^p(w)\, d\lambda(w).$$

Set $g(z) = f(z)/C_p$. Then $\int_{E(a,r)} g^p\, d\lambda \leq 1$, and

(10.2) $$g(z) \leq \frac{1}{s^{2n}} \int_{E(z,s)} g(w)\, d\lambda(w)$$

for all $z \in B$, and all s, $0 < s < 1$. Define A by

$$A = \sup\{g(z)^p (r - \epsilon)^{2n} : z \in E(a,\epsilon),\ 0 < \epsilon < r\}.$$

Since g is upper semicontinuous on B, A is finite. Let $z \in E(a,\epsilon)$, and let $\epsilon < s < r$. Then by (10.2),

$$g(z)(s - \epsilon)^{2n} \leq \int_{E(z,s-\epsilon)} g(w)\, d\lambda(w).$$

Since $\rho(z,w) = |\varphi_z(w)|$ is a metric on B (Lemma 8B.4),

$$E(z, s - \epsilon) \subset E(a, s) \subset E(a, r).$$

Therefore,

$$
\begin{aligned}
g(z)(s - \epsilon)^{2n} &\leq \int_{E(a,s)} g^p(w) g^{1-p}(w)\, d\lambda(w) \\
&\leq \left[\frac{A}{(r-s)^{2n}}\right]^{(1-p)/p} \int_{E(a,s)} g^p(w)\, d\lambda(w) \\
&\leq \left[\frac{A}{(r-s)^{2n}}\right]^{(1-p)/p}.
\end{aligned}
$$

Take $s = (\epsilon + r)/2$. Then $s - \epsilon = r - s = (r - \epsilon)/2$, and thus

$$g(z)(r - \epsilon)^{2n/p} \leq 2^{2n/p} A^{(1-p)/p}.$$

Therefore,

$$g^p(z)(r - \epsilon)^{2n} \leq 2^{2n} A^{1-p}.$$

Taking the supremum over $z \in E(a,r)$, $0 < \epsilon < r$, gives

$$A \leq 2^{2n} A^{1-p}, \qquad \text{or} \qquad A \leq 2^{2n/p}.$$

Therefore, $g^p(a)(r - \epsilon)^{2n} \leq 2^{2n/p}$ for all ϵ, $0 < \epsilon < r$. Thus

$$f^p(a) \leq \frac{2^{2n/p}}{r^{2n}} \int_{E(a,r)} f^p(w)\, d\lambda(w),$$

which is the desired inequality. $\square$

As a consequence of the proposition, if h is $\mathcal{M}$-harmonic on B, then for all p, $0 < p < \infty$,

(10.3) $$|h(a)|^p \leq \frac{C(n, p, r)}{r^{2n}} \int_{E(a,r)} |h(w)|^p\, d\lambda(w),$$

where $C(n, p, r)$ is given by (10.1). Our next goal is to obtain several inequalities for $|\tilde{\nabla} h(z)|$, where for a real valued C^1 function $|\tilde{\nabla} h|$ is as defined by (3.16).

Proposition 10.2. *Suppose h is $\mathcal{M}$-harmonic on B. For fixed δ, $0 < \delta < 1$, there exists a constant C_δ such that for all p, $0 < p < \infty$,*

$$|\tilde{\nabla} h(z)|^p \leq \frac{C_\delta}{\delta^{2n}} \int_{E(z,\delta)} |h(w)|^p\, d\lambda(w),$$

for all $z \in B$.

Proof. Let χ be a nonnegative C^∞ radial function with support contained in $\frac{\delta}{2} B$ and $\int_B \chi d\lambda = 1$. Then by Proposition 4.7,

$$h(z) = (h * \chi)(z) = \int_B \chi(\varphi_w(z)) h(w)\, d\lambda(w),$$

and as a consequence,

$$\tilde{\nabla} h(z) = \int_B \tilde{\nabla}_z \chi(\varphi_w(z)) h(w)\, d\lambda(w).$$

Thus, as in Theorem 9.3,

$$|\tilde{\nabla} h(z)| \leq \int_B |\tilde{\nabla}_z \chi(\varphi_w(z))| |h(w)|\, d\lambda(w) \leq C_\delta \sup_{w \in E(z,\frac{\delta}{2})} |h(w)|,$$

where

$$C_\delta = \int_{E(z,\frac{\delta}{2})} |\tilde{\nabla}_z \chi(\varphi_w(z))|\, d\lambda(w).$$

By the $\mathcal{M}$-invariance of $\tilde{\nabla}$,

$$|\tilde{\nabla}_z \chi(\varphi_w(z))| = |(\tilde{\nabla} \chi)(\varphi_w(z))| = |(\tilde{\nabla} \chi)(\varphi_z(w))|.$$

The last identity follows from the fact that $\tilde{\nabla}\chi$ is also radial on B. Therefore,

$$C_\delta = \int_{E(z,\frac{\delta}{2})} |(\tilde{\nabla}\chi)(\varphi_z(w))|\, d\lambda(w) = \int_{\frac{\delta}{2}B} |\tilde{\nabla}\chi(w)|\, d\lambda(w).$$

Thus the constant C_δ is independent of z. By Proposition 10.1,

$$|h(w)|^p \leq \frac{C(n,p,\delta)}{\delta^{2n}} \int_{E(w,\frac{\delta}{2})} |h|^p\, d\lambda.$$

If $w \in E(z,\frac{\delta}{2})$, then $E(w,\frac{\delta}{2}) \subset E(z,\delta)$. Therefore,

$$|\tilde{\nabla}h(z)|^p \leq \frac{C_\delta}{\delta^{2n}} \int_{E(z,\delta)} |h(w)|^p d\lambda(w),$$

for an appropriate constant C_δ. $\square$

If h is a nonnegative $\mathcal{M}$-harmonic function on B, then by the above $|\tilde{\nabla}h(z)| \leq C_\delta h(z)$. The following proposition sharpens this result.

Proposition 10.3. *If h is a nonnegative $\mathcal{M}$-harmonic function on B, then for all $z \in B$,*

$$|\tilde{\nabla}h(z)| \leq \frac{2n}{\sqrt{n+1}}\, h(z).$$

Furthermore, if $h(z) = \mathcal{P}[\nu](z)$, where ν is a finite signed measure on S, then

$$|\tilde{\nabla}h(z)| \leq \frac{2n}{\sqrt{n+1}}\mathcal{P}[|\nu|](z), \qquad z \in B$$

where $|\nu|$ is the total variation of ν.

Proof. If h is a nonnegative $\mathcal{M}$-harmonic function on B, then by Proposition 5.10, $h(z) = \mathcal{P}[\nu](z)$, where ν is a finite Borel measure on S. Thus

$$\tilde{\nabla}h(z) = \int_S \tilde{\nabla}_z\mathcal{P}(z,t)\, d\nu(t),$$

and as in the previous proposition,

$$|\tilde{\nabla}h(z)| \leq \int_S |\tilde{\nabla}_z\mathcal{P}(z,t)|\, d\nu(t).$$

Since $\mathcal{P}(z,t)$ is $\mathcal{M}$-harmonic on B, by (3.13) and (5.7),

$$|\tilde{\nabla}_z\mathcal{P}(z,t)|^2 = \frac{1}{2}\tilde{\Delta}\mathcal{P}^2(z,t) = \frac{4n^2}{n+1}\mathcal{P}^2(z,t).$$

Therefore,

$$|\tilde{\nabla}_z \mathcal{P}(z,t)| = \frac{2n}{\sqrt{n+1}} \mathcal{P}(z,t),$$

from which the first result now follows.

Suppose $h(z) = \mathcal{P}[\nu](z)$ where ν is a finite signed measure on S. Write $\nu = \nu^+ - \nu^-$, where ν^+ and ν^- are the positive and negative variations of ν respectively. Then

$$h(z) = h^+(z) - h^-(z),$$

where $h^+ = \mathcal{P}[\nu^+]$ and $h^- = \mathcal{P}[\nu^-]$. Therefore, by the above,

$$|\tilde{\nabla}h(z)| \leq |\tilde{\nabla}h^+(z)| + |\tilde{\nabla}h^-(z)| \leq \frac{2n}{\sqrt{n+1}}(\mathcal{P}[\nu^+](z) + \mathcal{P}[\nu^-](z)),$$

from which the result now follows. $\square$

We now turn to prove an analogue of Proposition 10.1 for $|\tilde{\nabla}h|$, for an $\mathcal{M}$-harmonic function h. One of the difficulties encountered is that $|\tilde{\nabla}h(z)|$ is in general not $\mathcal{M}$-subharmonic on B if h is $\mathcal{M}$-harmonic. To get around this difficulty we introduce the following vector fields on B. For $1 \leq j \leq n$, and h a C^1 function on B, set

$$X_j h(z) = \partial_j h(z) - \bar{z}_j \bar{R}h(z),$$

where

$$\partial_j h = \frac{\partial h}{\partial z_j}, \qquad \text{and} \qquad \bar{R}h(z) = \sum_{i=1}^{n} \bar{z}_i \frac{\partial h}{\partial \bar{z}_i}.$$

Although we will not require it, one can show that

$$X_j h(w) = \frac{\partial}{\partial z_j} h(\varphi_z(-w))\Big|_{z=0}.$$

This follows from the following identity, which is easily derived from (1.13):

$$(1 - \langle w, z \rangle)\varphi_z(w) = \left(1 - \frac{\langle w, z \rangle}{1 + s_z}\right) z - s_z w,$$

where $s_z = \sqrt{1 - |z|^2}$. One useful consequence of this identity is that the mapping $z \to \varphi_z(w)$ for fixed w is also C^∞ on B. The importance of the vector fields X_j are in the following:

Proposition 10.4. *If h is $\mathcal{M}$-harmonic on B, then $X_j h$ is $\mathcal{M}$-harmonic for all $j = 1, ..., n$.*

Proof. For convenience, set $a_{\alpha,\beta} = \delta_{\alpha,\beta} - \bar{z}_\alpha z_\beta$. To show $X_j h$ is $\mathcal{M}$-harmonic, it suffices to show that

$$\sum_{\alpha,\beta} a_{\alpha,\beta}(z) \frac{\partial^2 X_j h}{\partial z_\beta \partial \bar{z}_\alpha} = 0.$$

We first note that

$$\frac{\partial^2 X_j h}{\partial z_\beta \partial \bar{z}_\alpha} = -\frac{\partial}{\partial z_j} \left(\frac{\partial^2 h}{\partial z_\beta \partial \bar{z}_\alpha} \right) + \bar{z}_j \frac{\partial^2 h}{\partial z_\beta \partial \bar{z}_\alpha}$$

$$+ \bar{z}_j \sum_{i=1}^{n} \bar{z}_i \frac{\partial}{\partial \bar{z}_i} \left(\frac{\partial^2 h}{\partial z_\beta \partial \bar{z}_\alpha} \right) + \delta_{\alpha,j} \sum_{i=1}^{n} \bar{z}_i \frac{\partial^2 h}{\partial z_\beta \partial \bar{z}_i}.$$

Since h is $\mathcal{M}$-harmonic,

$$\sum_{\alpha,\beta} a_{\alpha,\beta}(z) \frac{\partial}{\partial z_j} \left(\frac{\partial^2 h}{\partial z_\beta \partial \bar{z}_\alpha} \right) = \sum_{\alpha,\beta} \delta_{\beta,j} \bar{z}_\alpha \frac{\partial^2 h}{\partial z_\beta \partial \bar{z}_\alpha} = \sum_{\alpha=1}^{n} \bar{z}_\alpha \frac{\partial^2 h}{\partial z_j \partial \bar{z}_\alpha},$$

and similarly

$$\sum_{\alpha,\beta} a_{\alpha,\beta}(z) \frac{\partial}{\partial \bar{z}_i} \left(\frac{\partial^2 h}{\partial z_\beta \partial \bar{z}_\alpha} \right) = \sum_{\beta=1}^{n} z_\beta \frac{\partial^2 h}{\partial z_\beta \partial \bar{z}_i}.$$

Therefore,

$$\sum_{\alpha,\beta} a_{\alpha,\beta}(z) \frac{\partial^2 X_j h}{\partial z_\beta \partial \bar{z}_\alpha} = -\sum_{\alpha,\beta} a_{\alpha,\beta}(z) \frac{\partial}{\partial z_j} \left(\frac{\partial^2 h}{\partial z_\beta \partial \bar{z}_\alpha} \right)$$

$$+ \bar{z}_j \sum_{i=1}^{n} \bar{z}_i \sum_{\alpha,\beta} a_{\alpha,\beta}(z) \frac{\partial}{\partial \bar{z}_i} \left(\frac{\partial^2 h}{\partial z_\beta \partial \bar{z}_\alpha} \right) + \sum_{i=1}^{n} \bar{z}_i \sum_{\alpha,\beta} a_{\alpha,\beta}(z) \delta_{\alpha,j} \frac{\partial^2 h}{\partial z_\beta \partial \bar{z}_i}$$

$$= -\sum_{\alpha=1}^{n} \bar{z}_\alpha \frac{\partial^2 h}{\partial z_j \partial \bar{z}_\alpha} + \bar{z}_j \sum_{i=1}^{n} \bar{z}_i \sum_{\beta=1}^{n} z_\beta \frac{\partial^2 h}{\partial z_\beta \partial \bar{z}_i} + \sum_{i=1}^{n} \bar{z}_i \sum_{\beta=1}^{n} a_{j,\beta}(z) \frac{\partial^2 h}{\partial z_\beta \partial \bar{z}_i}$$

$$= -\sum_{\alpha=1}^{n} \bar{z}_\alpha \frac{\partial^2 h}{\partial z_j \partial \bar{z}_\alpha} + \bar{z}_j \sum_{i,\beta} \bar{z}_i z_\beta \frac{\partial^2 h}{\partial z_\beta \partial \bar{z}_i} + \sum_{i=1}^{n} \bar{z}_i \frac{\partial^2 h}{\partial z_j \partial \bar{z}_i} - \bar{z}_j \sum_{i,\beta} \bar{z}_i z_\beta \frac{\partial^2 h}{\partial z_\beta \partial \bar{z}_i}$$

$$= 0. \qquad \square$$

As a consequence of Proposition 10.4, if h is $\mathcal{M}$-harmonic on B, then $|X_j h|^p$ is $\mathcal{M}$-subharmonic on B for all p, $1 \le p < \infty$. The following lemma makes the connection between $|\tilde{\nabla} h|$ and $|X_j h|$.

Lemma 10.5. *Let h be a C^2 real valued function on B. Then*

$$(10.4) \qquad |\tilde{\nabla}h(z)|^2 \le \frac{4}{n+1} \sum_{j=1}^{n} |X_j h(z)|^2 \le \frac{(1+|z|^2)^2}{(1-|z|^2)^2} |\tilde{\nabla}h(z)|^2.$$

As a consequence, if $0 < p < \infty$, then there exist positive constants C_1 and C_2, depending only on n and p, such that

$$(10.5) \qquad C_1 |\tilde{\nabla}h(z)|^p \le \sum_{j=1}^{n} |X_j h(z)|^p \le C_2 \frac{|\tilde{\nabla}h(z)|^p}{(1-|z|^2)^p}$$

Proof. For convenience, let $\partial h = (\partial_1 h, ..., \partial_n h)$. By definition,

$$|X_j h|^2 = |\partial_j h|^2 + |z_j|^2 |Rh|^2 - 2\mathrm{Re}\, z_j \partial_j h\, Rh.$$

Therefore

$$\sum_{j=1}^{n} |X_j h|^2 = |\partial h|^2 + |z|^2 |Rh|^2 - 2\,\mathrm{Re}\,(Rh)^2.$$

Since $-|Rh|^2 \le \mathrm{Re}\,(Rh)^2 \le |Rh|^2$,

$$\sum_{j=1}^{n} |X_j h|^2 \ge |\partial h|^2 + |z|^2 |Rh|^2 - 2|Rh|^2$$

$$= |\partial h|^2 - (1 - |z|^2)|Rh|^2 - |Rh|^2.$$

But

$$|Rh|^2 = \left| \sum_{j=1}^{n} z_j \partial_j h \right|^2 \le |z|^2 |\partial h|^2.$$

Therefore,

$$\sum_{j=1}^{n} |X_j|^2 \ge (1 - |z|^2) \left[|\partial h|^2 - |Rh|^2 \right].$$

Thus by (3.16),

$$|\tilde{\nabla}h| \le \frac{4}{n+1} \sum_{j=1}^{n} |X_j h|^2.$$

For the other half of inequality (10.4), as above,

$$\sum_{j=1}^{n} |X_j h|^2 \le |\partial h|^2 + |z|^2 ||Rh|^2 + 2|Rh|^2$$

$$\le |\partial h|^2 + |z|^4 |\partial h|^2 + 2|z|^2 |\partial h|^2 = (1 + |z|^2)^2 |\partial h|^2.$$

But,

$$|\tilde{\nabla}h|^2 \ge \frac{4}{n+1}(1 - |z|^2) \left[|\partial h|^2 - |z|^2 |\partial h|^2 \right] = \frac{4}{n+1}(1 - |z|^2)^2 |\partial h|^2,$$

from which inequality (10.4) follows. The proof of inequality (10.5) is now straight forward, and thus is omitted. $\square$

Proposition 10.6. *If h is $\mathcal{M}$-harmonic on B, then for all $a \in B$, $0 < r < 1$, and $0 < p < \infty$,*

$$|\tilde{\nabla}h(a)|^p \le C_2 \frac{C(n,p,r)}{r^{2n}(1-r^2)^p} \int_{E(a,r)} |\tilde{\nabla}h(w)|^p \, d\lambda(w),$$

where C_2 is a constant depending only on n and p, and $C(n,p,r)$ is given by (10.1).

Proof. By the $\mathcal{M}$-invariance of $\tilde{\nabla}$ and (10.5),

$$|\tilde{\nabla}h(a)|^p = |\tilde{\nabla}(h \circ \varphi_a)(0)|^p \le C_1 \sum_{j=1}^{n} |X_j(h \circ \varphi_a)(0)|^p,$$

which by Proposition 10.1 and (10.5)

$$\le C_1 \frac{C(n,p,r)}{r^{2n}} \int_{B_r} \sum_{j=1}^{n} |X_j(h \circ \varphi_a)(w)|^p \, d\lambda(w)$$

$$\le C_2 \frac{C(n,p,r)}{r^{2n}} \int_{B_r} \frac{|\tilde{\nabla}(h \circ \varphi_a)(w)|^p}{(1-|w|^2)^p} \, d\lambda(w)$$

$$\le C_2 \frac{C(n,p,r)}{r^{2n}(1-r^2)^p} \int_{B_r} |\tilde{\nabla}(h \circ \varphi_a)(w)|^2 \, d\lambda(w).$$

By the $\mathcal{M}$-invariance of λ, the last integral is equal to $\int_{E(a,r)} |\tilde{\nabla}h| \, d\lambda$. □

10.2. On a Theorem of Hardy and Littlewood.

In this section, we consider a generalization of a well known result of Hardy and Littlewood concerning the comparitive rate of growth of $M_p(f,r)$ and $M_p(f',r)$, where for $0 < p < \infty$ and $0 < r < 1$,

$$M_p(f,r) = \left[\int_S |f(rt)|^2 \, d\sigma(t) \right]^{1/p}.$$

As usual, $M_\infty(f,r) = \sup\{|f(rt)| : t \in S\}$. The classical result of Hardy and Littlewood [HL] is as follows: if f is holomorphic in U, $0 < p \le \infty$, and $\alpha > 0$, then

$$M_p(f,r) = O\left((1-r)^{-\alpha}\right) \iff M_p(f',r) = O\left((1-r)^{-(\alpha+1)}\right).$$

The extension to $\mathcal{M}$-harmonic functions on B and the invariant gradient is as follows:

Theorem 10.7. [Pa, Theorem 3.4] *If h is $\mathcal{M}$-harmonic on B, $0 < p \le \infty$, and $\alpha > 0$, then*

(10.6a) $\qquad M_p(h, r) = O\left(\dfrac{1}{(1 - r^2)^\alpha}\right),$ $\qquad$ as $\quad r \to 1$,

if and only if

(10.6b) $\qquad M_p(\widetilde{\nabla} h, r) = O\left(\dfrac{1}{(1 - r^2)^\alpha}\right),$ $\qquad$ as $\quad r \to 1$.

Remark: Although we stated the result for $\mathcal{M}$-harmonic functions, the result is also true for holomorphic functions on B. Furthermore, since

$$|\widetilde{\nabla}(\operatorname{Re} f)| = \sqrt{2}|\widetilde{\nabla} f|,$$

a holomorphic function f satisfies (10.6a) if and only if $\operatorname{Re} f$ satisfies (10.6a). Also, when $n = 1$, it should be noted that

$$M_p(\widetilde{\nabla} f, r) = (1 - r^2) M_p(f', r).$$

For the proof of the theorem we need the following lemma:

Lemma 10.8. *If f is a real valued C^1 function on B, and $\gamma : [a, b] \to B$ is a C^1 curve, then*

$$|f(\gamma(b)) - f(\gamma(a))| \le \sqrt{n+1} \int_a^b \frac{|\widetilde{\nabla} f(\gamma(t))| |\gamma'(t)|}{(1 - |\gamma(t)|^2)}\, dt.$$

Proof. Set $g(t) = f(\gamma(t))$. Then

$$|g'(t)| = 2\,|\operatorname{Re}\langle \partial f(\gamma(t)), \overline{\gamma'(t)}\rangle| \le 2|\partial f(\gamma(t))||\gamma'(t)|$$
$$\le \sqrt{n+1}\frac{|\widetilde{\nabla} f(\gamma(t))||\gamma'(t)|}{(1 - |\gamma(t)|^2)},$$

from which the result follows upon integration. $\quad\square$

Proof of Theorem 10.7. Suppose $M_p(h, r) = O((1 - r^2)^{-\alpha})$. Fix δ, $0 < \delta < 1$. Then by Proposition 10.2 and the $\mathcal{U}$-invariance of λ,

$$|\widetilde{\nabla} h(rUe_1)|^p \le C_\delta \int_{E(rUe_1, \delta)} |h(w)|^p\, d\lambda(w) = C_\delta \int_{E(re_1, \delta)} |h(Uw)|^p\, d\lambda(w)$$

for any $U \in \mathcal{U}$. Thus by (1.9),

$$M_p^p(\widetilde{\nabla} h, r) = \int_{\mathcal{U}} |\widetilde{\nabla} h(rUe_1)|^p \, dU \leq C_6 \int_{E(re_1, \delta)} \int_{\mathcal{U}} |h(Uw)|^p \, dU \, d\lambda(w)$$

$$= C_6 \int_{E(re_1, \delta)} M_p^p(h, |w|) d\lambda(w).$$

But if $w \in E(re_1, \delta)$, $(1 - |w|^2) \approx (1 - r^2)$, from which (10.6b) follows.

Conversely, suppose h satisfies (10.6b). Without loss of generality we may assume $h(0) = 0$. Then for $\zeta \in S$, $0 < r < 1$, by Lemma 10.8,

$$|h(r\zeta)| \leq C \int_0^r \frac{|\widetilde{\nabla} h(s\zeta)|}{(1 - s^2)} \, ds.$$

For $j = 0, 1, 2, \ldots$, set $r_j = 1 - 2^{-j}$. Also, for $0 < r < 1$, let m be such that $r_{m-1} < r \leq r_m$. Then by the above,

$$|h(r\zeta)| \leq C \sum_{j=0}^{m-1} \int_{r_j}^{r_{j+1}} \frac{|\widetilde{\nabla} h(s\zeta)|}{(1 - s^2)} \, ds$$

$$\leq C \sum_{j=0}^{m-1} \sup\{|\widetilde{\nabla} h(s\zeta)| : r_j < s < r_{j+1}\}.$$

If $r_j < s < r_{j+1}$, then

$$|\varphi_{r_j\zeta}(s\zeta)|^2 = \frac{(1 - r_j s)^2 - (1 - r_j^2)(1 - s^2)}{(1 - r_j s)^2} < \frac{(s - r_j)^2}{(1 - r_j)^2} < \frac{1}{4}.$$

Thus $s\zeta \in E(r_j\zeta, \frac{1}{2})$ for all s, $r_j < s < r_{j+1}$. As a consequence, $E(s\zeta, \frac{1}{4}) \subset E(r_j\zeta, \frac{3}{4})$ for all such s. Thus by Proposition 10.6 (with $r = \frac{1}{4}$),

$$\sup\{|\widetilde{\nabla} h(s\zeta)|^p : r_j < s < r_{j+1}\} \leq C \int_{E(r_j\zeta, \frac{3}{4})} |\widetilde{\nabla} h(w)|^p \, d\lambda(w).$$

Therefore, as in the first part of the proof,

$$\int_{\mathcal{U}} \sup\{|\widetilde{\nabla} h(sUe_1)| : r_j < s < r_{j+1}\} \, dU$$

$$\leq C \int_{\mathcal{U}} \int_{E(r_je_1, \frac{3}{4})} |\widetilde{\nabla} h(Uw)|^p \, d\lambda(w) \, dU$$

$$= C \int_{E(r_je_1, \frac{3}{4})} M_p(\widetilde{\nabla} h, |w|) \, d\lambda(w).$$

But by (10.6b), for $w \in E(r_j e_1, \frac{3}{4})$, $M_p^p(\tilde{\nabla} h, |w|) \leq C (2^j)^{p\alpha}$. Therefore,

$$M_p^p(h, r) \leq C \sum_{j=0}^{m-1} (2^{p\alpha})^j \leq C \frac{1}{(1 - r^2)^{p\alpha}},$$

which is the desired inequality. $\square$

Remark: The same proof shows that

$$M_p(h, r) = o \left((1 - r^2)^{-\alpha} \right) \iff M_p(\tilde{\nabla} h, r) = o \left((1 - r^2)^{-\alpha} \right).$$

The proof of the thereom also shows that if $M_p(h, r)$ is bounded, then so is $M_p(\tilde{\nabla} h, r)$. The converse however, as the following example illustrates, is false.

Example: ([Pa]) Let

$$h(z) = \frac{p}{n}(1 - z_1)^{-n/p}.$$

By Proposition 8.18, for all $p > 0$,

$$M_p^p(h, r) \geq c \log \frac{1}{(1 - r^2)}$$

for some positive constant c. On the other hand,

$$|\tilde{\nabla} h(z)|^2 = \frac{4}{n+1} \frac{(1 - |z|^2)(1 - |z_1|^2)}{|1 - z_1|^{2(\frac{n}{p}+1)}}.$$

Therefore, for $n \geq 2$,

$$M_p^p(\tilde{\nabla} h, r) = c_n (1 - r^2)^{p/2} \int_S \frac{(1 - r^2|t_1|^2)^{p/2}}{|1 - rt_1|^{n+p}} \, d\sigma(t),$$

which by identity (1.10),

$$= c_n (1 - r^2)^{p/2} \int_0^1 (1 - \rho^2)^{n-2} (1 - r^2\rho^2)^{p/2} \int_0^{2\pi} \frac{d\theta}{|1 - r\rho e^{i\theta}|^{n+p}} \, \rho \, d\rho.$$

By Proposition 8.18,

$$\int_0^{2\pi} \frac{d\theta}{|1 - r\rho e^{i\theta}|^{n+p}} \leq C (1 - r^2\rho^2)^{-n-p+1}.$$

Therefore, since $(1 - \rho^2) \leq (1 - r^2\rho^2)$,

$$M_p(\tilde{\nabla} h, r) \leq C (1 - r^2)^{p/2} \int_0^1 (1 - r^2\rho^2)^{-\frac{p}{2}-1} \rho \, d\rho \leq C \frac{1}{r^2}.$$

Thus $M_p(\tilde{\nabla} h, r)$ is bounded as $r \to 1$.

10.3. $\mathcal{M}$-Harmonic Bergman and Dirichlet Spaces.

In this final section, we take a brief look at the $\mathcal{M}$-harmonic Bergman and Dirichlet spaces on B. As for holomorphic functions, the $\mathcal{M}$-harmonic Bergman spaces are defined as follows:

Definition. *For $0 < p < \infty$, and $\gamma \in \mathbb{R}$, the weighted Bergman space $\mathcal{A}_p^\gamma$ is defined as the space of $\mathcal{M}$-harmonic functions h on B for which*

$$(10.7) \qquad \|h\|_{p,\gamma} = \left[\int_B (1 - |z|^2)^\gamma |h(z)|^p \, d\lambda(z) < \infty \right]^{1/p} < \infty.$$

Closely related to the $\mathcal{M}$-harmonic Bergman spaces are the weighted Dirichlet spaces, which are defined as follows:

Definition. *For $0 < p < \infty$, and $\gamma \in \mathbb{R}$, the $\mathcal{M}$-**harmonic Dirichlet space** $\mathcal{D}_p^\gamma$ is defined as the space of $\mathcal{M}$-harmonic functions h on B for which*

$$(10.8) \qquad \int_B |\tilde{\nabla} h(w)|^p (1 - |w|^2)^\gamma \, d\lambda(w) < \infty.$$

For $h \in \mathcal{D}_p^\gamma$, set

$$(10.9) \qquad |||h|||_{p,\gamma} = |h(0)| + \left(\int_B |\tilde{\nabla} h(w)|^p (1 - |w|^2)^\gamma \, d\lambda(w) \right)^{1/p}.$$

When $n = 1$, by identity (3.16),

$$|\tilde{\nabla} h(z)|^2 = 2(1 - |z|^2)^2 \left| \frac{\partial h}{\partial z} \right|^2 = \frac{1}{2}(1 - |z|^2)^2 \left(\left(\frac{\partial h}{\partial x} \right)^2 + \left(\frac{\partial h}{\partial y} \right)^2 \right),$$

and thus

$$(10.10) \qquad \int_U |\tilde{\nabla} h(z)|^p (1 - |z|^2)^\gamma \, d\lambda(z) = \frac{\sqrt{2}}{\pi} \int_U (1 - |z|^2)^{\gamma + p - 2} \left| \frac{\partial h}{\partial z} \right|^p \, dA(z),$$

where dA denotes area measure on U. The special case $\gamma = 0$, $p = 2$, gives the classical Dirichlet space on U. If f is holomorphic on U, then

$$|\tilde{\nabla} f(z)| = (1 = |z|^2)|f'(z)|,$$

and thus $f \in \mathcal{D}_p^\gamma(U)$ if and only if

$$\int_U (1 - |z|^2)^{\gamma + p - 2} |f'(z)|^p \, dA(z) < \infty.$$

Although we defined the spaces $\mathcal{D}_p^\gamma$ and $\mathcal{A}_p^\gamma$ for real valued $\mathcal{M}$-harmonic functions, these spaces can also be defined for holomorphic functions on B. However, since $|\tilde{\nabla}(\text{Re } f)| = \sqrt{2}|\tilde{\nabla} f|$ whenever f is holomorphic, we have $f \in \mathcal{D}_p^\gamma$ if and only if Re $f \in \mathcal{D}_p^\gamma$.

The spaces $\mathcal{D}_p^\gamma$ have been considered by several authors, both for holomorphic and $\mathcal{M}$-harmonic functions on B. In [MaS] these spaces were defined, for the special case $p = 2$, for holomorphic functions on B in terms of the magnitude of the euclidean gradient $|\partial f|$. In [AFJP], Arazy, Fisher, Janson and Peetre, considered the space $\mathcal{D}_p^0 \cap \text{Hol}(B)$ for the case $p > 2n$. One of the key results of their paper is that $f \in \mathcal{D}_p^0 \cap \text{Hol}(B)$, $p > 2n$, if and only if

$$(10.11) \qquad \int_B (1 - |z|^2)^p |\partial f(z)|^p \, d\lambda(z) < \infty.$$

In the same paper they also introduced the "diagonal" Besov spaces B_p^s, which for $-\infty < s < \infty$, $0 < p < \infty$, is defined as the space of holomorphic functions on B for which

$$(1 - |z|^2)^{m-s} R^m f(z) \in L^p \left(\frac{d\nu(z)}{(1 - |z|^2)} \right),$$

where m is any integer larger than s, and Rf is the radial derivative of f. These spaces have also been considered by Beatrous and Burbea [BB], and by M. Peloso [Pe].

For $\mathcal{M}$-harmonic functions, the spaces $\mathcal{D}_p^\gamma$ were considered by Hahn and Youssfi in [HY1, HY2] for the case $\gamma = 0$, $p > 2n$, and in [HY3] for other values of γ and p. In these papers these spaces were referred to as Besov spaces.

Our first result will be to show that for $p \geq 1$, the spaces $\mathcal{A}_p^\gamma$ and $\mathcal{D}_p^\gamma$ are trivial unless $\gamma > n$ and $\gamma > n - p$ respectively.

Proposition 10.9. *For all p, $1 \leq p < \infty$,*

(a) $\mathcal{A}_p^\gamma \neq \{0\}$ *if and only if $\gamma > n$, and*

(b) $\mathcal{D}_p^\gamma = \{\text{constants}\}$ *for all $\gamma \leq n - p$.*

Proof. (a) We first show that $\mathcal{A}_p^\gamma = \{0\}$ for all $\gamma \leq n$. Suppose $h \in \mathcal{A}_p^\gamma$. Let $0 < \rho < 1$ be arbitrary. Then

$$\int_{|z| \geq \rho} (1 - |z|^2)|h(z)|^p \, d\lambda$$

$$\geq 2n \int_\rho^{(1+\rho)/2} r^{2n-1}(1 - r^2)^{\gamma-n-1} \int_S |h(rt)|^p \, d\sigma(t) \, dr,$$

which since $|h|^p$ is $\mathcal{M}$-subharmonic,

$$\geq 2n \, M_p^p(h, \rho) \rho^{2n-1} \int_\rho^{(1+\rho)/2} (1 - r^2)^{\gamma-n-1} \, dr.$$

But for $\gamma \neq n$,

$$\int_\rho^{(1+\rho)/2} (1-r^2)^{\gamma-n-1}\, dr \geq C \int_\rho^{(1+\rho)/2} (1-r)^{\gamma-n-1}\, dr = C_1(1-\rho)^{\gamma-n},$$

where C_1 is a positive constant depending only and γ and n. Therefore,

$$(10.12) \qquad M_p^p(h,\rho) \leq \frac{C}{\rho^{2n-1}(1-\rho^2)^{\gamma-n}} \int_{|z|\geq\rho} (1-|z|^2)^\gamma |h(z)|^p\, d\lambda(z).$$

Thus if $\gamma < n$,

$$\lim_{\rho\to 1} M_p^p(h,\rho) = 0,$$

and as a consequence, $h \equiv 0$ on B. If $\gamma = n$, then

$$\int_\rho^{(1+\rho)/2} (1-r^2)^{-1}\, dr \geq \frac{1}{2} \int_\rho^{(1+\rho)/2} (1-r)^{-1}\, dr = \frac{1}{2}\log 2.$$

Therefore,

$$M_p^p(h,\rho) \leq C \int_{|z|\geq\rho} (1-|z|^2)^{-1}|h(z)|^p\, d\lambda(z).$$

If $\|h\|_{p,n} < \infty$, then the integral on the right converges to zero as $\rho \to 1$. Therefore, as above, $h \equiv 0$ on B.

If $\gamma > n$, then any bounded harmonic function is in $\mathcal{A}_p^\gamma$.

(b) Suppose $h \in \mathcal{D}_p^\gamma$, $p \geq 1$. By inequality (10.5),

$$|\widetilde{\nabla}h(z)|^p \geq C(1-|z|^2)^p \sum_{j=1}^n |X_j h(z)|^p.$$

Therefore,

$$\int_B (1-|z|^2)^{\gamma+p}|X_j h(z)|^p\, d\lambda(z) < \infty$$

for all $j = 1,...,n$. Since $|X_j h|^p$ is $\mathcal{M}$-subharmonic on B, as above

$$(10.13) \quad M_p^p(X_j h,\rho) \leq \frac{C(1-\rho^2)^{n-\gamma-p}}{\rho^{2n-1}} \int_{|z|\geq\rho} (1-|z|^2)^{\gamma+p}|X_j h(z)|^p\, d\lambda(z).$$

Therefore, as above, if $n-\gamma-p \geq 0$, $X_j h \equiv 0$ on B for all $j = 1,...,n$. This however is the case if and only if h is constant on B. $\quad\square$

Remarks:

(a) For all p, $0 < p < \infty$, $\mathcal{A}_p^\gamma \cap \text{Hol}(B) \neq \{0\}$ if and only if $\gamma > n$. However, as the following example will show, when $0 < p < 1$, $\mathcal{A}_p^\gamma$ will be nontrivial for some values of $\gamma < n$.

If f is holomorphic on B, then $|f(z)|^p$ is plurisubharmonic, and thus $\mathcal{M}$-subharmonic, for all p, $0 < p < \infty$. Therefore, as in the proof of (10.12), for $\gamma \neq n$,

$$M_p^p(f, \rho) \leq \frac{C}{\rho^{2n-1}(1-\rho^2)^{\gamma-n}} \int_{|z| \geq \rho} (1-|z|^2)^\gamma |f(z)|^p \, d\lambda(z).$$

Thus if $\gamma < n$, $\lim_{\rho \to 1} M_p^p(f, \rho) = 0$, and as a consequence $f \equiv 0$ on B. The proof of the case $\gamma = n$ follows similarly.

(b) For all p, $0 < p < \infty$, $\mathcal{D}_p^\gamma \cap \text{Hol}(B) = \{constants\}$ for all $\gamma \leq n - p$.

The proof of (b) follows as in the proof of Proposition 10.9, replacing $|X_j h|^p$ by $|\frac{\partial h}{\partial z_j}|^p$, which is also plurisubharmonic for all p, $0 < p < \infty$.

(c) When $n = 1$, as a consequence of (10.10), $h(z) = z \in \mathcal{D}_p^\gamma \cap \text{Hol}(B)$ for all $\gamma > 1 - p$. Thus $\mathcal{D}_p^\gamma$ is nontrivial for all γ satisfying $\gamma + p > 1$. When $n > 1$, it is easily shown that $\mathcal{D}_p^\gamma$ contains nonconstant functions for all $\gamma > n - \frac{p}{2}$. For example,

$$h(z) = z_1 + \bar{z}_1 \in \mathcal{D}_p^\gamma \quad \text{for all} \quad \gamma > n - \frac{p}{2}.$$

When $n \geq 2$, it was shown by Hahn and Yousffi [HY2], that $\mathcal{D}_p^0 \cap \text{Hol}(B)$ is nontrivial if and only if $p > 2n$. It is conjectured that for $n \geq 2$, $\mathcal{D}_p^\gamma \cap \text{Hol}(B)$ is nontrivial if and only if $\gamma > n - \frac{p}{2}$.

(d) Even though the spaces $\mathcal{A}_p^\gamma$ and $\mathcal{D}_p^\gamma$ are nontrivial for some $\gamma < n$ when $p < 1$, one can easily show that for any $p > 0$, $\mathcal{A}_p^\gamma = \{0\}$ whenever $\gamma \leq 0$. Similarly, for any $p > 0$, $\mathcal{D}_p^\gamma$ is trivial whenever $\gamma \leq -p$.

Example: This example shows that for $0 < p < 1$, the space $\mathcal{A}_p^\gamma$ contains nonzero functions for some values of $\gamma < n$. Let $h(z) = \mathcal{P}(z, e_1)$. Then

$$\int_B (1-|z|^2)^\gamma |h(z)|^p \, d\lambda(z)$$

$$= 2n \int_0^1 r^{2n-1}(1-r^2)^{\gamma+np-n-1} \int_S \frac{d\sigma(t)}{|1-\langle rt, e_1 \rangle|^{2np}} \, dr.$$

By Proposition 8.18,

$$\int_S \frac{d\sigma(t)}{|1-\langle rt, e_1 \rangle|^{2np}} \leq C \begin{cases} (1-r^2)^{n-2np}, & \frac{1}{2} < p < 1, \\ \log \frac{1}{(1-r^2)}, & p = \frac{1}{2}, \\ 1, & 0 < p < \frac{1}{2}. \end{cases}$$

From this it now follows that for $0 < p < 1$, $\mathcal{P}(z, e_1) \in \mathcal{A}_p^\gamma$ for all γ satisfying

$$\gamma > np, \qquad \frac{1}{2} \leq p < 1,$$

$$\gamma > n(1-p), \qquad 0 < p < \frac{1}{2}.$$

Thus for $0 < p < 1$, the spaces $\mathcal{A}_p^\gamma$ are nontrivial for some values of $\gamma < n$. Since

$$|\tilde{\nabla}\mathcal{P}(z,t)| = \frac{2n}{\sqrt{n+1}}\mathcal{P}(z,t),$$

the same result is true for $\mathcal{D}_p^\gamma$ whenever $p < 1$.

We next consider the relationship between the spaces $\mathcal{A}_p^\gamma$ and $\mathcal{D}_p^\gamma$.

Theorem 10.10. *Let h be $\mathcal{M}$-harmonic on B.*

(a) For all p, $0 < p < \infty$, and $\gamma \in \mathbb{R}$, there exists a constant C, independent of h, such that

$$(10.14) \qquad \int_B (1-|z|^2)^\gamma |\tilde{\nabla}h(z)|^p \, d\lambda(z) \le C \int_B (1-|z|^2)^\gamma |h(z)|^p \, d\lambda(z).$$

(b) For all p, $1 \le p < \infty$, and $\gamma > n$, there exists a positive constant C, independent of h, such that

$$(10.15) \quad C \int_B (1-|z|^2)^\gamma |h(z)|^p \, d\lambda(z) \le |h(0)|^p + \int_B (1-|z|^2)^\gamma |\tilde{\nabla}h(z)|^p \, d\lambda(z).$$

Corollary 10.11. *For $p \ge 1$, and $\gamma > n$, $h \in \mathcal{D}_p^\gamma$ if and only if $h \in \mathcal{A}_p^\gamma$ with*

$$C\,\|h\|_{p,\gamma} \le |||h|||_{p,\gamma} \le C\,\|h\|_{p,\gamma}$$

for some positive constant C.

Proof. (a) Let $0 < p < \infty$, and fix δ, $0 < \delta < 1$. Then as in the proof of Theorem 10.7,

$$M_p^p(\tilde{\nabla}h, r) \le C_\delta \int_{E(re_1,\delta)} M_p^p(h, |w|) \, d\lambda(w).$$

Therefore,

$$\int_B (1-|z|^2)^\gamma |\tilde{\nabla}h(z)|^p \, d\lambda(z) = \int_B (1-|z|^2)^\gamma M_p^p(\tilde{\nabla}h, |z|) \, d\lambda(z)$$

$$\le C_\delta \int_B (1-|z|^2)^\gamma \int_B \chi_{E(|z|e_1,\delta)}(w) M_p^p(h, |w|) \, d\lambda(w) d\lambda(z),$$

which by Fubini's theorem,

$$= \int_B M_p^p(h, |w|) \left[\int_B \chi_{E(|z|e_1,\delta)}(1-|z|^2)^\gamma \, d\lambda(z) \right] d\lambda(w).$$

Since $(1 - |z|^2) \approx (1 - |w|^2)$ for $w \in E(|z|e_1, \delta)$,

$$\int_B \chi_{E(|z|e_1,\delta)}(w)(1 - |z|^2)^\gamma \, d\lambda(z) \leq c_\delta (1 - |w|^2)^\gamma \lambda(E(w, \delta)).$$

Therefore, since $\lambda(E(w, \delta))$ depends only on δ,

$$\int_B (1 - |z|^2)^\gamma |\widetilde{\nabla} h(z)|^p \leq C_\delta \int_B (1 - |w|^2)^\gamma M_p^p(h, |w|) \, d\lambda(w),$$

which establishes (10.14).

(b) Suppose $p \geq 1$ and $\gamma > n$. Let $\zeta \in S$ and $0 < r < 1$. Then by Lemma 10.8 with $\gamma(t) = t\zeta$,

$$|h(r\zeta)| \leq |h(0)| + C \int_0^r \frac{|\widetilde{\nabla} h(t\zeta)|}{(1 - t^2)} \, dt.$$

If $r \leq \frac{1}{2}$, then by Hölder's inequality,

$$M_p^p(h, r) \leq C_1 |h(0)|^p + C_2 \int_0^r M_p^p(\widetilde{\nabla} h, t) \, dt.$$

By inequality (10.5),

$$\int_0^r M_p^p(\widetilde{\nabla} h, t) \, dt \leq C \sum_{j=1}^n \int_0^r M_p^p(X_j h, t) \, dt$$

which since $|X_j h|^p$ is $\mathcal{M}$-subharmonic, and $r \leq \frac{1}{2}$,

$$\leq \sum_{j=1}^n M_p^p(X_j h, r) \leq C \, M_p^p(\widetilde{\nabla} h, r).$$

Therefore,

$$(10.16) \quad \int_{\frac{1}{2}B} (1 - |z|^2)^\gamma |h(z)|^p \leq C_1 |h(0)|^p + C_2 \int_{\frac{1}{2}B} (1 - |z|^2)^\gamma |\widetilde{\nabla} h(z)|^p \, d\lambda(z).$$

We now consider the integral over $\frac{1}{2} \leq |z| < 1$. Since $\gamma > n$, choose $\alpha > 0$ such that $\gamma - \alpha p > n$. For $p > 1$, let $q = p/(p-1)$ be the conjugate exponent of p. Then by Hölder's inequality,

$$\int_0^r \frac{|\widetilde{\nabla} h(t\zeta)|}{(1 - t^2)} \, dt \leq \left[\int_0^r (1 - t^2)^{-\alpha q - 1} \, dt \right]^{1/q} \left[\int_0^r (1 - t^2)^{\alpha p - 1} |\widetilde{\nabla} h(t\zeta)|^2 \right]^{1/p}$$

$$\leq C_{\alpha,q} (1 - r^2)^{-\alpha} \left[\int_0^r (1 - t^2)^{\alpha p - 1} |\widetilde{\nabla} h(t\zeta)|^p \, dt \right]^{1/p}.$$

Therefore,

$$\int_S |h(r\zeta)|^p \, d\sigma(\zeta) \le 2^p |h(0)|^p + C \, (1-r^2)^{-\alpha p} \int_0^r (1-t^2)^{\alpha p - 1} M_p^p(\tilde{\nabla} h, t) \, dt.$$

It is easily seen that this inequality is still valid when $p = 1$. Thus since $\gamma > n$,

$$\int_{|z| > \frac{1}{2}} (1 - |z|^2)^\gamma |h(z)|^p \, d\lambda(z) \le C_1 |h(0)|^p + C_2 (I_1 + I_2),$$

where

$$I_1 = \int_{\frac{1}{2}}^1 r^{2n-1} (1-r^2)^{\gamma - n - \alpha p - 1} \left[\int_0^{\frac{1}{2}} (1-t^2)^{\alpha p - 1} M_p^p(\tilde{\nabla} h, t) \, dt \right] dr,$$

and

$$I_2 = \int_{\frac{1}{2}}^1 r^{2n-1} (1-r^2)^{\gamma - n - \alpha p - 1} \left[\int_{\frac{1}{2}}^r (1-t^2)^{\alpha p - 1} M_p^p(\tilde{\nabla} h, t) \, dt \right] dr.$$

For the first integral,

$$I_1 \le C \int_{r = \frac{1}{2}}^1 \int_{t=0}^{\frac{1}{2}} r^{2n-1} (1-r^2)^{\gamma - n - 1} M_p^p(\tilde{\nabla} h, t) \, dt \, dr \le C \, M_p^p(\tilde{\nabla} h, \tfrac{1}{2}),$$

which as in the proof of Proposition 10.9(b),

$$\le \int_{|z| > \frac{1}{2}} (1 - |z|^2)^\gamma |h(z)|^p \, d\lambda(z).$$

For the second integral, since

$$\{(r,t) : \tfrac{1}{2} \le r < 1, \ \tfrac{1}{2} \le t \le r\} = \{(r,t) : \tfrac{1}{2} \le t < 1, \ t \le r < 1\},$$

by interchanging the order of integration,

$$I_2 = \int_{\frac{1}{2}}^1 (1-t^2)^{\alpha p - 1} M_p^p(\tilde{\nabla} h, t) \left[\int_t^1 r^{2n-1} (1-r^2)^{\gamma - n - \alpha p - 1} \, dr \right] dt,$$

which since $\gamma - \alpha p - n > 0$,

$$\le C \int_{\frac{1}{2}}^1 (1-t^2)^{\gamma - n - 1} M_p^p(\tilde{\nabla} h, t) \, dt$$

$$\le C \int_{|z| > \frac{1}{2}} (1 - |z|^2)^\gamma |\tilde{\nabla} h(z)|^p \, d\lambda(z).$$

Therefore,

$$\int_{|z|>\frac{1}{2}} (1 - |z|^2)^\gamma |h(z)|^p \, d\lambda(z)$$

$$\leq C_1 |h(0)|^p + C_2 \int_{|z|>\frac{1}{2}} (1 - |z|^2)^\gamma |\widetilde{\nabla} h(z)|^p \, d\lambda(z),$$

which when combined with (10.16) proves (10.15). $\square$

Note: Part (a) of Theorem 10.10 is due to Pavlovic([Pa, Theorem 3.1]). In that paper, the author also proves that the reverse inequality holds for all p and γ whenever h is a nonzero eigenfunction of $\widetilde{\Delta}$.

10.4. Remarks.

We close this chapter by indicating some relationships between the $\mathcal{M}$-harmonic Hardy spaces $\mathcal{H}^p$ and the spaces $\mathcal{A}_p^\gamma$ and $\mathcal{D}_p^\gamma$. As in Section 5.3, for $0 < p < \infty$, $\mathcal{H}^p$ denotes the set of $\mathcal{M}$-harmonic functions h on B for which

$$\|h\|_p^p = \sup_{0<r<1} \int_S |h(rt)|^p \, d\sigma(t) < \infty.$$

By Theorem 6.18, for $1 < p < \infty$, $h \in \mathcal{H}^p$ if and only if

$$\int_B (1 - |z|^2)^n |h(z)|^{p-2} |\widetilde{\nabla} h(z)|^2 \, d\lambda(z) < \infty.$$

Thus when $p = 2$, $\mathcal{H}^2 = \mathcal{D}_2^n$.

(1) For all p, $2 \leq p < \infty$, $\mathcal{H}^p \subset \mathcal{D}_p^n$, with

$$|||h|||_{p,n} \leq C_{n,p} \|h\|_p, \qquad \text{for all} \quad h \in \mathcal{H}^p,$$

where $C_{n,p}$ is a constant depending only on n and p.

Since equality holds when $p = 2$, we assume $2 < p < \infty$. Suppose $h \in \mathcal{H}^p$ is nonnegative. Then by Proposition 10.3, $|\widetilde{\nabla} h(z)| \leq c_n h(z)$ for all $z \in B$. Therefore since $p - 2 > 0$,

$$\int_B (1 - |z|^2)^n |\widetilde{\nabla} h(z)|^p \, d\lambda(z) = \int_B (1 - |z|^2)^n |\widetilde{\nabla} h(z)|^{p-2} |\widetilde{\nabla} h(z)|^2 \, d\lambda(z)$$

$$\leq C_n \int_B (1 - |z|^2)^n h(z)^{p-2} |\widetilde{\nabla} h(z)|^2 \, d\lambda(z),$$

from which the result follows by Theorem 6.18. Since every $h \in \mathcal{H}^p$ can be written as $h = h_1 - h_2$, where h_1, h_2 are positive harmonic functions in $\mathcal{H}^p$, the result follows by linearity.

(2) For all p, $1 \leq p \leq 2$, $\mathcal{D}_p^n \subset \mathcal{H}^p$.

As in (7.23), for $\alpha > 1$, $\zeta \in S$, let

$$(10.17) \qquad S_\alpha f(\zeta) = \left(\int_{D_\alpha(\zeta)} |\widetilde{\nabla} f|^2 \, d\lambda \right)^{1/2}$$

denote the area integral of f. To prove (2), we first establish the following:

Lemma 10.12. *Let $0 < p \leq 2$, $\alpha > 1$, and $\zeta \in S$. Then there exists a finite constant C, independent of α and ζ, such that*

$$S_\alpha f(\zeta) \leq C \left(\int_{D_\beta(\zeta)} |\tilde\nabla f|^p \, d\lambda \right)^{1/p},$$

for all $\mathcal{M}$-harmonic functions f on B.

Proof. Let $E(z) = \varphi_z(B_{\frac{1}{2}})$, and suppose f is $\mathcal{M}$-harmonic on B. By Proposition 10.6, for all $z \in B$,

$$|\tilde\nabla f(z)|^2 \leq C \left(\int_{E(z)} |\tilde\nabla f(w)^p| \, d\lambda(w) \right)^{2/p}.$$

By Lemma 8.11, $E(z) \subset D_\beta(\zeta)$ for all $z \in D_\alpha(\zeta)$ and any $\beta \geq 3\alpha$. Therefore,

$$S_\alpha f(\zeta) \leq C \left(\int_{D_\alpha(\zeta)} \left(\int_{D_\beta(\zeta)} \chi_{E(z)}(w) |\tilde\nabla f(w)|^p \, d\lambda(w) \right)^{2/p} \right)^{1/2},$$

which for $0 < p \leq 2$, by Jessen's inequality

$$\leq C \left(\int_{D_\beta(\zeta)} \left(\int_{D_\alpha(\zeta)} \chi_{E(w)}(z) |\tilde\nabla f(w)|^2 \, d\lambda(z) \right)^{p/2} d\lambda(w) \right)^{1/p}$$

$$\leq C \left(\int_{D_\beta(\zeta)} |\tilde\nabla f(w)|^p \lambda(E(w)) \, d\lambda(w) \right)^{1/p} \leq C \left(\int_{D_\beta(\zeta)} |\tilde\nabla f|^p \, d\lambda \right)^{1/p}.$$

$\square$

As a consequence of the lemma,

$$\int_S [S_\alpha f(\zeta)]^p \, d\sigma(\zeta) \leq C \int_S \int_{D_\beta(\zeta)} |\tilde\nabla f(w)|^p \, d\lambda(w) \, d\sigma(\zeta),$$

which by Lemma 8.12

$$\leq C \int_B (1 - |w|^2)^n |\tilde\nabla f(w)|^p \, d\lambda(w).$$

Thus $S_\alpha f \in L^p(S)$ for all p, $0 < p \leq 2$. By Theorem 5.1 of [Ge], this implies that $f \in H^p$ for all p, $1 \leq p \leq 2$.

(3) Another consequence of Lemma 10.12 is the following: if $f \in \mathcal{D}_p^n$, $0 < p \leq 2$, then f has admissible limits a.e. on S. As above, if $f \in \mathcal{D}_p^n$, $0 < p \leq 2$, then $S_\alpha f \in L^p(S)$, and thus is finite a.e. on S. By Theorem 3.1 of [Ge], this implies that f has admissible limits a.e. on S.

(4) If $\gamma < n$, $1 < p < \infty$, then $\mathcal{D}_p^\gamma \subset \mathcal{H}^p$.

This is easily shown as follows: for $\frac{1}{2} < r < 1$, and $\zeta \in S$, by Lemma 10.8,

$$|h(r\zeta)| \leq |h(\tfrac{1}{2}\zeta)| + C \int_{\frac{1}{2}}^{1} \frac{|\tilde{\nabla} h(t\zeta)|}{(1-t^2)} \, dt,$$

which by Hölder's inequality,

$$\leq |h(\tfrac{1}{2}\zeta)| + C \left[\int_{\frac{1}{2}}^{r} (1-t^2)^{\gamma-n-1} |\tilde{\nabla} h(t\zeta)|^p \right]^{1/p}.$$

From this it now follows that $h \in \mathcal{H}^p$.

(5) If $f \in \mathcal{D}_p^\gamma$ for some $\gamma < n$, then in analogy with (10.17) one can define the tangential area integral of f as follows: for $1 \leq \tau \leq n/\gamma$, $c > 0$, $\zeta \in S$, set

$$(10.18) \qquad S_{\tau,c} f(\zeta) = \left(\int_{T_{\tau,c}(\zeta)} |\tilde{\nabla} f|^2 \, d\lambda \right)^{1/2},$$

where $T_{\tau,c}(\zeta)$ is the tangential approach region as defined in (8.24). As in Lemma 10.12, if $f \in \mathcal{D}_p^\gamma$, $0 < p \leq 2$, $0 < \gamma < n$,

$$(10.19) \qquad S_{\tau,c} f(\zeta) \leq C \left(\int_{T_{\tau,c'}(\zeta)} |\tilde{\nabla} f|^p \, d\lambda \right)^{1/p},$$

for some $c' > c$, and for $1 \leq \tau \leq n/\gamma$,

$$(10.20) \qquad \int_S [S_{\tau,c} f(\zeta)]^p \, d\sigma(\zeta) \leq C \int_B (1-|w|^2)^{n/\tau} |\tilde{\nabla} f(w)|^p \, d\lambda(w).$$

It is conjectured that if $f \in \mathcal{D}_p^\gamma$, $0 \leq p \leq 2$, $0 < \gamma < n$, then f has T_τ-limits a.e. on S for all τ, $1 \leq \tau \leq n/\gamma$.

(6) The space $\mathcal{A}_1^\gamma$ plays a special role in the study of the holomorphic Hardy space $H^p = \mathcal{H}^p \cap \mathrm{Hol}(B)$, for $0 < p < 1$. As a consequence of Theorem 4 of [MiH],

$$H^p \subset \mathcal{A}_1^{n/p} \qquad \text{for all} \quad 0 < p < 1.$$

The space $\mathcal{A}_1^{n/p} \cap \text{Hol}(B)$, $0 < p < 1$ is usually denoted by B^p, and has been shown to be the containing Frechet space of H^p.

(7) When $n = 1$, $\mathcal{D}_2^\gamma \cap \text{Hol}(U)$ has a particularly nice characterization in terms of the Taylor coefficients of f. Suppose

$$f(z) = \sum_{k=0}^{\infty} a_k z^k$$

is holomorphic in U. Then $f \in \mathcal{D}_2^\gamma$ if and only if

$$\int_U (1 - |z|^2)^\gamma |f'(z)|^2 \, dA(z) < \infty.$$

But by orthogonality,

$$\int_U (1 - |z|^2)^\gamma |f'(z)|^2 \, dA(z) = \frac{1}{\pi} \sum_{k=1}^{\infty} k^2 |a_k|^2 \int_0^1 (1 - r^2)^\gamma r^{2k-2} \, r \, dr$$

$$= \frac{\Gamma(\gamma + 1)}{2\pi} \sum_{k=1}^{\infty} \frac{k^2 \Gamma(k)}{\Gamma(k + \gamma + 1)} |a_k|^2.$$

But

$$\frac{\Gamma(k)}{\Gamma(k + \gamma + 1)} \approx \frac{1}{k^{\gamma+1}}.$$

Therefore, $f \in \mathcal{D}_2^\gamma$ if and only if

$$\sum_{k=1}^{\infty} k^{1-\gamma} |a_k|^2 < \infty.$$

References

[Ah]   P. R. Ahern, *The mean modulus and the derivative of an inner function*, Indiana Univ. Math. J. **28** (1979), 311–347.

[ACl]   _____ and D. N. Clark, *On Inner functions with B^p derivative*, Michigan Math. J. **23** (1976), 107–118.

[ACo]   _____ and W. Cohn, *Exceptional sets for Hardy Sobolev functions, $p > 1$*, Indiana U. Math. J. **38** (1989), 417–453.

[AFR]   _____, M. Flores and W. Rudin, *An invariant volume mean value property*, J. Funct. Analysis **111** (1993), 380–397.

[Ai]   H. Aikawa, *Tangential behavior of Green potentials and contractive properties of L^p-potentials*, Tokyo J. Math. **9** (1986), 223–245.

[AFJP]   J. Arazy, S. Fisher, S. Janson, and J. Peetre, *Membership of Hankel operators on the ball in unitary ideals*, J. London Math. Soc. **43** (1991), 485–508.

[AH]   M. Arsove and A. Huber, *On the existence of non-tangential limits of subharmonic functions*, J. London Math. Soc. **42** (1967), 125–132.

[BB]   F. Beatrous and J. Burbea, *Holomorphic Sobolev spaces on the ball*, Dissertationes Math. **270** (1989), 1–57.

[Be]   S. Bergman, *The Kernel Function and Conformal Mapping*, Amer. Math. Soc., Providence, R.I., 1970.

[BC]   Robert D. Berman and W. S. Cohn, *Tangential limits of Blaschke products and functions of bounded mean oscillation*, Illinois J. of Math. **31** (1987), 218–239.

[BL]   A. Bonami and N. Lohoué, *Noyaux de Szegö de certains domanies de $\mathbb{C}^n$ et inégalités L^p*, C.R. Acad. Sci. Paris **285** (1977), 699–702.

[BT]   E. Bedford and B. A. Taylor, *The Dirichlet problem for the complex Monge-Ampere equation*, Invent. Math. **37** (1976), 1–44.

[Ca]   E. Cartan, *Sur les domaines bornés homogenes de l' espace de n variables complexes*, Abh. Math. Sem. Univ. Hamburg **11** (1935), 116–162.

[Car]   G. T. Cargo, *Angular and tangential limits of Blaschke products and their successive derviatives*, Can. J. Math. **14** (1962), 334–348.

[Ce]   U. Cegrell, *Capacities in Complex Analysis; Aspects of Mathematics*, vol. 14, Vieweg, Germany, 1988.

[CiS]   J. A. Cima and C. S. Stanton, *Admissible limits of M-subharmonic functions*, Michigan Math. J. **32** (1985), 211–220.

[Co] W. S. Cohn, *Non-isotropic Hausdorff measure and exceptional sets for holomorphic Sobolev functions*, Illinois J. Math. **33** (1989), 673–690.

[DA] J.P. D'Angelo, *A note on the Bergman Kernel*, Duke Math. J. **45** (1978), 259–265.

[De] J-P Demailly, *Mesures de Monge-Ampere et mesures pluriharmoniques*, Math. Z. **194** (1987), 519–564.

[Fa] P. Fatou, *Séries trigonometriques et séries de Taylor*, Acta Math. **30** (1906), 335–400.

[Fu1] H. Furstenberg, *A Poisson formula for semisimple Lie groups*, Ann. of Math. **77** (1963), 335–386.

[Fu2] ———, *Boundaries of Riemannian symmetric spaces; Symmetric Spaces*, Marcel Dekker, Inc., New York, N.Y., 1972.

[Ga] S. J. Gardiner, *Growth properties of pth means of potentials in the unit ball*, Proc. Amer. Math. Soc. **103** (1988), 861–869.

[Ge] D. Geller, *Some results in H^p theory for the Heisenberg group*, Duke Math. J. **47** (1980), 365–390.

[Gi] S.G. Gindikin, *Analysis in homogeneous domains*, Russian Math. Surveys **19** (1964), 1–89.

[Go] R. Godement, *Une généralization du théoréme de la moyenne pour les fonctions harmoniques*, C. R. Acad. Sci., Paris **234** (1952), 2137–2139.

[Ha] K.T. Hahn, *Properties of holomorphic functions of bounded characteristic on star-shaped circular domains*, J. Reine Angew. Math. **254** (1972), 33–40.

[HM1] ——— and J. Mitchell, *Green's function on the classical Cartan domains*, MRC Technical Summary Report No. 800 (1967).

[HM2] ——— and J. Mitchell, *H^p spaces on bounded symmetric domains*, Trans. Amer. Math. Soc. **146** (1969), 521–531.

[HSi] ——— and David Singman, *Boundary behavior of invariant Green's potentials on the unit ball in $\mathbb{C}^n$*, Trans. Amer. Math. Soc. **309** (1988), 339–354.

[HSt] ——— and M. Stoll, *Boundary limits of Green's potentials on the unit ball in $\mathbb{C}^n$*, Complex Variable **9** (1988), 359–371.

[HSY] ———, M. Stoll and E. H. Yousffi, *Invariant potentials and tangential boundary behavior of $\mathcal{M}$-subharmonic functions*, (preprint).

[HY1] ——— and E. H. Yousffi, *M-harmonic Besov p-spaces and Hankel operators in the Bergman space on the ball in $\mathbb{C}^n$*, Manuscripta Math. **71** (1991), 67–81.

[HY2] ——— and E. H. Yousffi, *Möbius invariant Besov p-spaces and Hankel operators on the ball in $\mathbb{C}^n$*, Complex Var. **17** (1991), 89–104.

[HY3]   ———— and E. H. Yousffi, *Tangential boundary behavior of M-harmonic Besov functions*, J. Math. Anal. Appl. **175** (1993), 206–221.

[HL]   G. H. Hardy and J. E. Littlewood, *Some properties of conjugate functions*, J. Reine Angew. Math. **167** (1931), 405–423.

[Hei]   M. Heins, *The minimum modulus of a bounded analytic function*, Duke Math. J. **14** (1947), 179–215.

[He1]   Sigurdur Helgason, *Differential Geometry, Lie Groups, and Symmetric Spaces*, Academic Press, New York, NY, 1978.

[He2]   ————, *Groups and Geometric Analysis*, Academic Press, New York, NY, 1984.

[Hel]   L. L. Helms, *Introduction to Potential Theory*, Wiley-Interscience, 1969.

[Ho]   L. Hörmander, *Linear Partial Differential Operators*, Springer-Verlag, New York, N.Y., 1963.

[Hu]   L. K. Hua, *Harmonic Analysis of Functions in Several Complex Variables in the Classical Domains*, Amer. Math. Soc., Providence, R.I., 1963.

[Ka]   F. I. Karpelevic, *The geometry of geodesics and the eigenfunctions of the Beltrami-Laplace operator on symmetric spaces*, Trans. Moscow Math. Soc. **14** (1965), 51-199.

[Ki]   J. R. Kinney, *Boundary behavior of Blaschke products in the unit circle*, Proc. Amer. Math. Soc. **12** (1961), 484–488.

[Kl1]   M. Klimek, *Extremal plurisubharmonic functions and invariant pseudo distances*, Bull. Soc. Math. Fr. **113** (1985), 123–142.

[Kl2]   ————, *Infinitesimal pseudo-metrics and the Schwarz lemma*, Proc. Amer. Math. Soc. **105** (1989), 134–140..

[Ko1]   A. Koranyi, *The Poisson integral for generalized half-planes and bounded symmetric domains*, Ann. Math. **82** (1965), 332-350.

[Ko2]   ————, *Harmonic functions on Hermitian hyperbolic space*, Trans. Amer. Math. Soc. **135** (1969), 507–516.

[Ko3]   ————, *Harmonic functions on symmetric spaces; Symmetric Spaces*, Marcel Decker, Inc., New York, N.Y., 1972.

[Ko4]   ————, *A survey of harmonic functions on symmetric spaces*, Proceedings of Symposia in Pure Mathematics, vol. XXXV, Amer. Math. Soc., Providence, R.I., 1979.

[Kr1]   S. G. Krantz, *Function Theory in Several Complex Variables*, John Wiley & Sons, New York, N.Y., 1982.

[Kr2]   ———, *Partial Differential Equations and Complex Analysis*, CRC Press, Inc., Boca Raton, Florida, 1992.

[Li]   J. E. Littlewood, *On functions subharmonic in a circle. III*, Proc. London Math. Soc. **32** (1931), 222-234.

[Lu]   D. H. Luecking, *Boundary behavior of Green potentials*, Proc. Amer. Math. Soc. **96** (1986), 481–488.

[MaS]   B. MacCluer and J. Shapiro, *Angular derivatives and compact composition operators on the Hardy and Bergman spaces*, Can. J. Math **38** (1986), 878–906.

[Mic]   H. L. Michelson, *Generalized Poisson integrals and their boundary behavior*, Amer. Math. Soc., Proceedings of Symposia in Pure Math. **26** (1973), 329–333.

[MiH]   J. Mitchell and K.T. Hahn, *Representation of linear functions in H^p spaces over bounded symmetric domains*, J. Math. Analysis & Appl. **56** (1976), 379–396.

[Miz]   Y. Mizuta, *On the existence of weighted tangential boundary limits of Green potentials in a half space*, (preprint).

[NR]   A.W. Nagel and W. Rudin, *Moebius-invariant function spaces on balls and spheres*, Duke Math. J. **43** (1976), 841-865.

[NRS]   A. W. Nagel, W. Rudin, and J. H. Shapiro, *Tangential boundary behavior of Dirichlet- type spaces*, Annals of Math. **116** (1982), 331-360.

[NS]   W.C. Nestlerode and M. Stoll, *Radial limits of n-subharmonic functions in the polydisc*, Trans. Amer. Math. Soc. **279** (1983), 691-703.

[Pa]   M. Pavlovic, *Inequalities for the gradient of eigenfunctions of the invariant laplacian in the unit ball*, Indag. Mathem., N.S. **2** (1991), 89–98.

[Pe]   M. Peloso, *Möbius invariant spaces on the unit ball*, Michigan Math. J. **39** (1992), 509–536.

[Pu]   R. Putz, *A generalized area theorem for harmonic functions on hermitian hyperbolic space*, Trans. Amer. Math. Soc. **168** (1972), 243–258.

[RaU]   W. Ramey and D. Ullrich, *The pointwise Fatou theorem and its converse for positive pluriharmonic functions*, Duke Math. J. **49** (1982), 655–675.

[Ri]   P. J. Rippon, *On the boundary behaviour of Green potentials*, Proc. London Math. Soc. **38** (1979), 461-480.

[Ru1]   W. Rudin, *Function Theory in Polydiscs*, W.A. Benjamin, Inc, New York, 1969.

[Ru2]   ———, *Pluriharmonic functions in balls*, Proc. Amer. Math. Soc. **62** (1977), 44–46.

[Ru3]   ———, *Function Theory in the Unit Ball of $\mathbb{C}^n$*, Springer-Verlag, New York, 1980.

[Ru4]   _____, *New Constructions of Functions Holomorphic in the Unit Ball of* $\mathbb{C}^n$, Amer. Math. Soc., Providence, R.I., 1986.

[Sj]   Peter Sjögren, *Fatou theorems and maximal functions for eigenfunctions of the Laplace-Beltrami operator in a bidisk*, J.Reine Angew. Math. **345** (1983), 93–110.

[Ste1]   E. M. Stein, *Singular Integrals and Differentiability Properties of Functions*, Princeton University Press, Princeton, New Jersey, 1970.

[Ste2]   _____, *Boundary Behavior of Holomorphic Functions of Several Complex Variables*, Mathematical Notes, Princeton University Press, Princeton, New Jersey, 1972.

[Ste3]   _____, *Boundary behavior of harmonic functions on symmetric spaces: maximal estimates for Poisson integrals*, Invent. Math. **74** (1983), 63–83.

[SW]   _____ and N. Weiss, *On the convergence of Poisson integrals*, Trans. Amer. Math. Soc. **140** (1969), 35–54.

[Sto1]   M. Stoll, *Integral formulae for pluriharmonic functions on bounded symmetric domains*, Duke Math. J. **41** (1974), 393–404.

[Sto2]   _____, *Hardy type spaces of harmonic functions on symmetric spaces of non-compact type*, J. Reine Angew. Math. **271** (1974), 63–76.

[Sto3]   _____, *Harmonic majorants for plurisubharmonic functions on bounded symmetric domains with applications to the spaces H_ϕ and N_**, J. Reine Angew. Math. **282** (1976), 80–87.

[Sto4]   _____, *Boundary limits of Green potentials in the unit disc*, Arch. Math. **56** (1985), 451-455.

[Sto5]   _____, *Rate of growth of p'th means of invariant potentials in the unit ball of* $\mathbb{C}^n$, J. Math. Analysis & Appl. **143** (1989), 480–499.

[Sto6]   _____, *Rate of growth of p'th means of invariant potentials in the unit ball of* $\mathbb{C}^n$, *II*, J. Math. Analysis & Appl. **165** (1992), 374–398.

[Sto7]   _____, *Tangential boundary limits of invariant potentials in the unit ball of* $\mathbb{C}^n$, J. Math. Analysis & Appl **177** (1993), 553–571.

[Sto8]   _____, *A characterization of Hardy spaces on the unit ball of* $\mathbb{C}^n$, J. London Math. Soc. **48** (1993), 126–136.

[Sto9]   _____, *Non-isotropic Hausdorff capacity of exceptional sets of invariant potentials*, Pot. Analysis (to appear).

[Ul1]   D. Ullrich, *Möbius-invariant Potential Theory in the unit ball of* $\mathbb{C}^n$, Ph.D. Dissertation, University of Wisconsin (1981).

[Ul2]   _____, *Radial limits of M-subharmonic functions*, Trans. Amer. Math. Soc. **292** (1985), 501–518.

[We]   N. J. Weiss, *Convergence of Poisson integrals on generalized upper half-planes*, Trans. Amer. Math. Soc. **136** (1969), 109–123.

[Wu]   Jang-Mei G. Wu, L^p-*densities and boundary behavior of Green potentials*, Indiana Univ. Math. J. **28** (1979), 895–911.

[Ya]   S. Yamashita, *Criteria for a function to be of Hardy class H^p*, Proc. Amer. Math. Soc. **75** (1979), 69–72.

[Zi]   L. Ziomek, *On the boundary behavior in the metric L^p of subharmonic functions*, Studia Math. **29** (1967), 97–105.

[Zy]   A. Zygmund, *Trigonometric Series, I & II*, Cambridge University Press, London, 1968.

Index

Admissible:
   domain, 82
   maximal function, 89
Admissible limits:
   in L^p, 120
   of Poisson integrals, 82
   of potentials, 107
Ahern, P., 42
Approximate identity, 36
Arazy, J., 154
Area integral, 94, 160
Arsove, M., 107
Automorphism, 9
   for ball, 11
   for polydisc, 41

Beatrous, F., 154
Bergman kernel, 13
   for ball, 16
   for polydisc, 18
Bergman metric, 20
   for ball, 22
Bergman space, 6, 153
Besov spaces, 154
Biholomorphic, 9
Blaschke product, 113, 124
Bonami, A., 44
Boundary measure, 64
Bounded symmetric domain, 2, 40
Burbea, J., 154

Cargo, G.T., 124
Cauchy-Szegö kernel, 44
Cima, J., 107, 120
Convolution, 34

Demailly, J.P., 79
Dirichlet problem:
   for ball, 48
   for rB, 56

Dirichlet space, 6, 153

Fatou, P., 81
Fatou's theorem, 81, 95
Fisher, J., 154
Flores, M., 42
Function:
   biholomorphic, 9
   harmonic, 1
   holomorphic, 8
   invariant harmonic, 3, 4, 31
   $\mathcal{M}$-harmonic, 4, 31
   $\mathcal{M}$-subharmonic, 31
   $\mathcal{M}$-superharmonic, 31
   n-harmonic, 2, 24
   n-subharmonic, 41
   pluriharmonic, 2, 9
   plurisubharmonic, 9
   radial, 26
Furstenberg, H., 3, 53, 56, 58

Gardiner, S.J., 125
Garnett, J.P., 51
Geller, D., 95
Gindiken, S. G., 44
Godement, R., 40
Gradient, 27
Graham, C.R., 51
Green's formula, 23
Green's function, 60
   for invariant laplacian, 65
   pluricomplex, 79
Green potential:
   of measure, 71
   of function, 107

Hahn, K.T., 65, 125, 126, 154
Hardy, G.H., 149
Hardy space, 52, 75

Harmonic:
  euclidean, 1
  invariant, 3, 4, 31
  $\mathcal{M}$-harmonic, 4, 31
  n-harmonic, 2, 24
  pluriharmonic, 2, 9
  strongly, 3, 40
  weakly, 3, 31
Harnack's inequality, 61
Heins, M., 124
Holomorphic function, 8
Homogeneous domain, 2
Hörmander, L., 56
Hua, L. K., 5, 44
Huber, A., 107
Hyperconvex, 79
Hypoadmissible, 93
Hyporadial, 93

Invariant:
  convolution, 34
  gradient, 27
  harmonic, 3, 4, 31
  laplacian, 1, 23
  measure, 19
  potential, 71, 74
  Riesz kernel, 133
  Riesz operator, 132
  Riesz potential, 127
  subharmonic, 31
Isotropic, 45

Jansen, S., 154

Karpelic, F. I., 58
Koranyi, A., 44, 53, 82
Krantz, S., 48, 51

Laplacian, 1
Laplace-Beltrami operator, 1, 23
  for ball, 25
  for polydisc, 4, 24
  radial form, 26
Least $\mathcal{M}$-harmonic majorant, 61
Li, S-Y., 79
Limits:
  admissible, 82

in L^p, 120
  tangential, 114
Littlewood, J.E., 149
Littlewood's theorem, 96
Locally integrable, 33
Lohoué, N., 44

Marcinkiewicz, J., 81
Maximal function:
  of measures, 83
  radial, 104
Maximum principle, 33
Mean value inequality, 31, 142
$\mathcal{M}$-harmonic:
  function, 4, 31
  majorant, 61
Michelson, H. L., 58
Mitchel, J., 58, 65
Möbius group, 11
Monge-Ampere operator, 80
$\mathcal{M}$-subharmonic, 31
$\mathcal{M}$-superharmonic, 31

n-harmonic, 2, 24
Noneuclidean disc, 34
Nonisotropic:
  ball, 46
  capacity, 123
  metric, 46, 85
Nontangential limit, 81
n-subharmonic, 41, 58

Orthogonal group, 10

Pavlovic, M., 160
Peetre, J., 154
Peloso, M., 154
Pluriharmonic, 2, 9
Plurisubharmonic, 9
Poisson kernel, 44, 45
Poisson-Szegö:
  integral, 51
  kernel, 44
Potential, 71, 74
Putz, R., 95

Radial function, 26

Radial limit:
  of Poisson integrals, 92
  of potentials, 96
Radial maximal function, 104
Reinhardt set, 15
Riesz:
  decomposition theorem, 70
  kernel, 126, 133
  measure, 40
  operator, 132
  potential, 127
Rudin, W., 42

Shur's theorem, 139
Sjögren, P., 95
Stanton, C. S., 107, 120
Stein, E. M., 81, 95
Strongly harmonic, 3, 40
Subharmonic:
  invariant, 31
  $\mathcal{M}$-subharmonic, 31
  n-subharmonic, 41, 58
  plurisubharmonic, 9
  strongly, 41
Support of a measure, 127
Symmetric domain, 2
Szegö kernel, 44

Tangential:
  area integral, 162
  boundary limit, 114
  region, 114

Ullrich, D., 53, 61, 65
Unitary group, 10
Upper derivate, 88

Weakly harmonic, 3, 31
Weakly subharmonic, 31
Weiss, N. J., 81
Wu, Jang-Wei, 115

Yamashita, S., 75
Yousffi, E. H., 126, 154

Ziomek, L. 120
Zygmund, A., 81

Printed in the United States
By Bookmasters